T0177021

The Biology of Small Mammals

JOSEPH F. MERRITT

The Johns Hopkins University Press

Baltimore

© 2010 The Johns Hopkins University Press
All rights reserved. Published 2010
Printed in the United States of America on acid-free paper
9 8 7 6 5 4 3 2

The Johns Hopkins University Press
2715 North Charles Street
Baltimore, Maryland 21218-4363
www.press.jhu.edu

Library of Congress Cataloging-in-Publication Data

Merritt, Joseph F.
    The biology of small mammals / Joseph F. Merritt.
        p.   cm.
    Includes bibliographical references and index.
    ISBN-13: 978-0-8018-7950-0 (hardcover : alk. paper)
    ISBN-10: 0-8018-7950-7 (hardcover : alk. paper)
    1. Mammals.   I. Title.
QL703.M47 2009
599—dc22   2009018389

A catalog record for this book is available from the British Library.

Special discounts are available for bulk purchases of this book. For more information, please contact Special Sales at 410-516-6936 or specialsales@press.jhu.edu.

The Johns Hopkins University Press uses environmentally friendly book materials, including recycled text paper that is composed of at least 30 percent post-consumer waste, whenever possible. All of our book papers are acid-free, and our jackets and covers are printed on paper with recycled content.

For Samantha

Contents

....

Preface

....

This book is dedicated to the understanding of small mammals, of their fascinating adaptations that enhance survival in many different environments. I have written this book to appeal to a broad readership, including nature lovers, wildlife professionals, and students of science. I have attempted to elucidate the great diversity of morphological, physiological, and behavioral adaptations that enable small mammals to cope with the many selection pressures present in their environments. While I discuss a broad range of small mammals, from shrews to carnivores, this book is by no means comprehensive. Rather, I highlight those species and adaptations I personally have found most interesting and enjoyable to learn, teach, and write about.

Adaptations are indicative of how animals make a living and are crucial in tailoring an individual for survival and enhancing its ability to reproduce. The diversity of adaptations exhibited in small mammals is tremendous and spans taxonomic borders. Adaptations range from the brown adipose tissue of masked shrews, used as a crucial heat-producing source in the cold, to the sanguinivorous (blood-eating) and piscivorous (fish-eating) behavior exhibited by vampire bats and bulldog bats, the bizarre acrobatic maneuvers of eastern spotted skunks, and monogamy and housekeeping in sengis.

Formalized by Charles Darwin in 1859, the term **adaptation** is a central concept in biology and has attracted a great deal of controversy. A discussion of this controversy is well beyond the scope of this book. I refer the interested reader to the thorough accounts of the history and meaning of "adaptation" provided by R. W. Hill and Wyse (1989) and Willmer, Stone, and Johnston (2000). I use the term here to describe any anatomical, physiological, or behavioral trait or character (or suite of characters) that results from natural selection and enhances the survivorship and reproductive capacity of a given species in its natural environment. I focus on delineating survival mechanisms with an emphasis on form and function and the adaptive significance of unique characters; however, it is noteworthy that some species show fascinating morphologies with no adaptive purpose that we know of.

I begin by answering the question, what is a small mammal? The majority of extant species of mammals are indeed small, measuring less than a meter and ranging in size from the diminutive Kitti's hog-nosed bat and Etruscan shrew, weighing less than 2 g (0.07 oz), to the familiar woodchuck weighing in at about 5 kg (11 lb)—this is where I draw the line. In Chapter 1, I trace the history of the definition of "small mammal" and discuss the advantages and disadvantages of being small. I then endeavor to define the protagonists: the small mammals. This is a formidable task and one that is subject to many interpretations and disagreements—all debated with good humor. To set the stage for the book, I present an overview of the cast of characters organized by order. I have organized the volume into three parts: Modes of Feeding, Environmental Adaptations, and Reproduction. Some species are given a bit more attention than others, and I highlight some such groups in Case Studies. My unequal allocation of attention among groups is simply due to my curiosity and interest in particular species, as well as my hope that the reader will share in this fascination.

Chapters 2 through 5 are dedicated to a discussion of modes of feeding. I review the great variation in feeding designs of small mammals, including insectivory, herbivory, carnivory, and omnivory. Foraging techniques reflect dietary opportunities and, as one would expect, these chapters have many players, and there are, of course, many more than can be included here. I hope I have covered most of the foraging strategies but realize that I have only scratched the surface in terms of practitioners. As case studies, I include (in Chapter 3) the fascinating "rock badgers" of Africa with their bizarre elastic foot pads and a one-of-a-kind digestive system and (in Chapter 5) brief discussions of dietary nonconformists such as the insect-eating bat-eared foxes and the most carnivorous North American mouse, the grasshopper mouse. And (also in chapter 5) I could not forget to discuss the hero shrew, or armored shrew, of Africa, that mysterious insectivore with its bizarre, fortified backbone for which we have difficulty ascribing any functional significance. I also take time to discuss the different foraging strategies of some of the familiar and not-so-familiar species of bats.

In Chapters 6 through 10, I discuss environmental adaptations, specifically the many physiological, morphological, and behavioral mechanisms adaptive in coping with the rigors of cold and hot environments. I survey the diverse mechanisms that small mammals employ to avoid and resist cold and heat and aridity. Avoidance techniques include the long, thinly furred ears of African fennecs, huddling thermogenesis in muskrats, and torpidity in Etruscan shrews. Resistance strategies include the role of brown adipose tissue in shrews, bats, and voles. In the case studies (Chapter 10), I discuss how some small mammals, such as the secretive southern flying squirrels, are especially adept at using group nesting for conserving heat in winter, and how hedgehogs employ hibernation to cope with the cold.

In Chapters 11 and 12, I discuss the reproductive patterns of small mammals, ranging from delayed development in New World fruit bats and delayed implantation in weasels to embryonic diapause in the smallest gliding mammals, the feather-tailed gliders of Australia. I also discuss the adaptive significance of the fastidious road maintenance and mate guarding of the monogamous African sengis, the absentee maternal care of treeshrews, the role of lek behavior in those "flying loudspeakers," the hammer-headed fruit bats of Africa, and the natural history of the only eusocial mammal, the African mole-rat. A fitting conclusion to this discussion of reproduction, and to the book, is Chapter 13, dedicated to a discussion (albeit far too brief) of the long-term research on the population cycles of lemmings and snowshoe hares.

As I mentioned earlier, I have written this book to appeal to a broad readership, including amateur naturalists, students of science, and wildlife professionals. I use scientific vocabulary throughout the book. Technical terms are shown in boldface type when they are first introduced in a chapter, and those terms are defined in the text and in the glossary. I provide up-to-date, as well as some historically interesting, reference sources throughout the chapters. General information on mammalian ordinal characteristics is derived from Feldhamer et al.'s *Mammalogy: Adaptation, Diversity, Ecology* (2007). The reader is referred to that volume for a more detailed description of taxa and characteristics of mammals. To further assist the reader in understanding the lives of small mammals and tracking mammalogy in general, the appendix includes a list of useful web sites.

Acknowledgments

....

The underlying motivation that inspired me to write this book began in my early childhood—with an innocuous gift from my parents of a pair of guinea pigs. This starter colony, of two presumed males, challenged my abilities in animal husbandry by resulting in a thriving and rather noisy colony of 50—almost overnight! This was the beginning of my infatuation with and interest in small mammals. And it was an incentive to learn more, which I have attempted to do in preparing this book. Small mammals bring great happiness to so many, and I hope this book will make some headway in furthering our understanding of and, most importantly, respect for this special group of mammals.

My son, Jeff, was instrumental in helping me craft the proposal for *Biology of Small Mammals*. He offered many helpful pointers on the most effective and interesting way of presenting the many diverse adaptations of small mammals to my readership—I hope I have succeeded in accomplishing this goal. Specific sections of the book benefited from comments and criticisms provided by Ron Barry, Ken Catania, Sara Churchfield, George Feldhamer, Fritz Geiser, Al Kurta, Eileen Lacey, Galen Rathbun, Michael Steele, Steven Vessey, and Jane Waterman. I appreciate their time, energy, and comments, which have greatly improved this book. I offer special thanks to Samantha Carpenter for proofreading the entire manuscript and for editing and correcting many sections. This book has benefited from Samantha's writing ability; her talents have greatly enhanced the clarity and understanding of the book.

I am especially grateful to Vincent Burke, Senior Editor of the Johns Hopkins University Press, for his guidance and patience with Merritt's inability to adhere to deadlines. When receiving a request for extension on a deadline (actually, more than one), Vince was always positive and understanding, with comments such as, "Okay, I will give you another month . . . Can you please write every day? That makes all the difference." I also thank Edwina (Winnie) Rodgers, Acquisitions Assistant at the Press, for her never-failing willingness to help with my myriad of questions. I extend my sincere thanks and admiration to Linda Strange for her expert copy editing of this book. Working with Linda has been a pleasure.

I am indebted to the following persons for providing images and illustrations, free of charge: David M. Armstrong, Cheryl Asa, Stan Braude, Hynek Burda, Kenneth Catania, Joseph Cook, Dennis Cullinane, Jerry W. Dragoo, Fritz Geiser, Ken Geluso, Jill Gordon, Werner Haberl, Hendrik Hoeck, Susan Hoffman, Rimvydas Juskaitis, Asko Kaikusalo, Michael B. Kingston, Gerhard Koertner, Hal Korber, Rexford Lord, Robert MacArthur, Michael W. McKelvey, Peter Meserve, Russell A. Mittermeier, Luboš Mráz, Nathan Muchhala, Phillip Myers, Michael J. O'Farrell, James Parnell, Polly Phillips, Galen Rathbun, Peggy G. Rismiller, Kevin Schafer, Peter Schouten, Chris and Mathilde Stuart, Andrey Tchabovsky, Richard Thorington, Øivind Tøien, Peter Vogel, Jane M. Waterman, Howard P. Whidden, Pat Woolley, and Heather York. Ronald Nowak graciously granted permission for me to use selected images from his book *Walker's Mammals of the World*. I also thank an anonymous reviewer for providing a thorough review of the manuscript.

The production of this book also drew upon the talents of Carie Nixon of the Illinois Natural History Survey, who prepared several superb spot illustrations. Artists are often seen as tough to get along with, but Carie clearly departs from this stereotype. Beth Wohlgemuth and Susan Braxton of the Library of the Illinois Natural History Survey were always eager and ready to assist me with tracking down hard-to-find research titles. I offer special thanks to Ryan Rehmeier for his assistance with securing permissions for images from the ASM (American Society of Mammalogists) image library; Ryan was a pleasure to work with. Kelly Paralis Keenen, owner of Penumbra Design, Inc., graciously supplied the Johns Hopkins University Press with files of some 20 illustrations from the third edition of *Mammalogy: Adaptation, Diversity, Ecology* (Feldhamer et al. 2007). Last, I thank my colleagues Joyce Hofmann and Jean Mengelkoch, of the Illinois Natural History Survey, for tolerating my preoccupation with the writing of this book.

The Biology of Small Mammals

ONE

Introduction

....

What Is a Small Mammal?
....

There are 5,416 currently recognized, extant species of mammals throughout the world (Wilson and Reeder 2005). Most of these mammals are "small"
in total length (head and body), measuring less than 12 inches (about 31 cm),
and in weight, with about 90% of all mammals weighing less than 5 kg
(11 lb); most are rodents, bats, and shrews. Many definitions of the term
"small mammal" have been proposed. In *The Ecology of Small Mammals*,
Michael Delany (1974) defined "small mammals" as including "insectivores" and rodents no larger than 120 g (4.2 oz), whereas Heusner (1991)
assigned an upper limit of 20 kg (44 lb)! However, most authors have settled on a 5 kg limit. For example, Bourliere (1975: 2), in a chapter in *Small
Mammals: Their Productivity and Population Dynamics* (Golley, Petrusewicz, and Ryszkowski 1975), set the upper limit for a small mammal at
5 kg, and in *Ecology of Small Mammals*, Stoddart (1979) settled on the
5 kg size limit. Most recently, Allan Degen (1997) also adopted the 5 kg
limit in his book *Ecophysiology of Small Desert Mammals*. Dr. William H.
Burt, eminent mammalogist and long-time Curator of Mammals of the University of Michigan Museum of Zoology, touched on this issue during my
Ph.D. comprehensive exam at the University of Colorado. Dr. Burt asked
me to define a small mammal and, as I recall, the correct answer involved
something about the size of a bread box!

In this book, I side with the majority and define a small mammal as a
mammal weighing less than 5 kg, thus placing the woodchuck (*Marmota
monax*), weighing in at about 5 kg, as the largest small mammal. Terms
such as "large," "medium," and "small" are, of course, arbitrary and rela-

tive. Most mammals discussed in this book would clearly be considered small; however, I occasionally deviate slightly and discuss certain species that are not commonly thought of as the stereotypical small mammal—for example, some of the small primates, canids and felids, platypuses, and echidnas. However, these groups do have species that fall under the 5 kg limit and, I believe, exhibit some fascinating and unique **adaptations** for survival, and thus are worthy of discussion in this book.

Advantages and Disadvantages of Being a Small Mammal

The advantages and disadvantages of small size are outlined by Bourliere (1975). Ted Fleming (1979) also discusses this topic in his chapter "Life History Strategies" in Stoddart's book. I summarize their comments here.

One obvious advantage of smallness in the natural environment pertains to easy concealment from predators: by hiding rather than fleeing, a small mammal can cut back on energy expended for locomotion—a crucial economy for hot-bodied small mammals. Most species of small mammals are **nocturnal** or **crepuscular** (active at dawn and dusk) and therefore are faced with predation pressure from aerial predators. Foraging below the leaf litter, under the snow (**subnivean**), or in the **subterranean** environment certainly reduces vulnerability to capture from above. For small mammals such as "insectivores," rodents, and small marsupials, concealing coat coloration adds an additional advantage to small size in avoiding detection from aerial predators. Also, small size affords a wider range of available food types. Small mammals focus their foraging on the availability of resources to meet their energy needs. Some small mammals, such as mice, consume almost everything they come in contact with, including invertebrates, plant material such as rhizomes, fruits, seeds, leaves, flowers, and ferns, and fungi.

Small size also enables small mammals to utilize microhabitats that are well insulated from fluctuating temperatures in the ambient environment. Small bodies are heat sinks, due to their high surface-area-to-volume ratios: they must conserve energy by reducing conductive heat loss. By living in subterranean tunnels under forest floors, below the grass cover of prairies, or, better yet, in subnivean runways below a blanket of snow, small mammals are able to reduce heat loss in spite of fluctuating ambient temperatures. Small size enables these mammals to utilize microhabitats characterized by ameliorating microclimatic temperatures and humidities. Another advantage of being a small mammal is the high fertility rate seen in many species. Most small mammals have short life spans, short generation times, and high turnover rates; as a consequence, changes in the genetic composition of a population may occur in a short time. Such mechanisms provide populations of small mammals with resiliency and the possibility of adapting to and coping with habitat changes.

What about the disadvantages or constraints of being small? Most species of small mammals, as noted above, have short life spans. For instance, shrews do not live more than 12 months in the wild, and most voles and mice in the temperate zones rarely make it through two winters. Bats are an exception and tend to be long-lived for their size (Altringham 1996). Mortality in bats is high in the first year of life and decreases rapidly over the next few years. Bats are known to live 10 to 30 years. Brandt's myotis (*Myotis brandti*), found in Siberia, is reported to live up to 41 years in the wild. For the majority of small mammals, however, short life spans prevent elaborate socialization of young and thus limit opportunities for learning from the experiences of adults. Such experiences are crucial for refining food-procuring abilities and avoiding predators—to mention just a couple of learned behaviors. Although small mammals do have an enhanced availability of food supplies in the wild, their rate of intake is great: many species of shrews are known to consume more than the equivalent of their own body mass each day. However, the major disadvantages to being small are related to the high energy cost of **euthermy** (maintaining normal body temperature). These challenges are detailed in Chapters 6 through 9.

In sum, the advantages of small size outweigh the disadvantages. In this book, I discuss the many varied adaptations of small mammals that make up their suite of adaptive responses to the diverse environmental selective pressures that have accumulated over evolutionary time.

Small mammals (and mammals in general) are newcomers on the evolutionary stage. They evolved from a primitive group of reptiles (the therapsids) in the Triassic Period of the Mesozoic Era, only some 230 million years ago, during the "Age of Reptiles." At the end of the Cretaceous Period, about 65 million years ago, the ruling reptiles nearly became extinct. The mammals then underwent a remarkable adaptive radiation. This led to the dominant position of mammals in the Cenozoic Era. Over the past 65 million years, a time aptly called the "Age of Mammals," these animals have diversified into a great variety of ecological niches. Of the 29 orders of mammals currently recognized (Wilson and Reeder 2005), 13 contain mostly small mammals. Of the 5,416 recognized species of mammals, more than 90% weigh less than 5 kg. Small mammals are represented in each of the three major mammalian lineages, the prototherians (monotremes), the metatherians (marsupials), and the eutherians (placentals). They are found in aquatic environments, underground, and in the air, as well as in trees and on the ground. Small mammals range from northern polar regions to the tropics on all continents except Antarctica. Although well represented and, without a doubt, successful, small mammals face many challenges. So exactly who are the small mammals that are the subjects of this volume? A brief discussion of the representative orders and families will help put things in perspective.

The Protagonists

The class Mammalia is divided into two subclasses of extant taxa, based primarily on reproductive characteristics: the Prototheria and the Theria. Prototherians are a small group of four or five living species in the order Monotremata. Therians make up the vast majority of mammals. They are divided into two infraclasses: the metatherians and the eutherians. Metatherians are more commonly referred to as **marsupials**; eutherians are typically called **placental** mammals. The approximately 330 extant species of marsupials, along with the monotremes, make up about 6% of the world's mammalian fauna. The remainder is made up of eutherians. In the following overview, I highlight only those orders that contain mammals weighing less than 5 kg. In this book, I follow the classification of Wilson and Reeder (2005). The following general information on mammalian ordinal characteristics is derived from Feldhamer et al. (2007).

Monotremes and Marsupials

Mammals evolved from synapsid (reptilian) ancestors over a long time period. Monotremes (subclass Prototheria) differ significantly from marsupials and eutherians (subclass Theria) in their retention of various reptilian features (Feldhamer et al. 2007). Extant monotremes are represented by two families comprising five species. The family Ornithorhynchidae is monotypic (a group that includes a single taxon) and includes only the duck-billed platypus (*Ornithorhynchus anatinus*). The family Tachyglossidae includes the short-billed echidna (*Tachyglossus aculeatus*) and long-billed echidna (*Zaglossus bruijni*). Two additional species of long-billed echidnas in New Guinea, *Z. bartoni* and *Z. attenboroughi,* were recognized by Flannery and Groves (1998) based on characteristics of the claws and cranium, with the latter species described from a single specimen. Both families of monotremes contain members that weigh less than 5 kg. On an evolutionary time scale, the monotremes probably owe their continued survival to relative geographic isolation from eutherians.

All marsupials were once placed in a single order, the Marsupialia. About 6.5% of all species of mammals are marsupials. In this book, I follow recent taxonomy by Wilson and Reeder (2005) that recognizes seven orders of marsupials comprising 331 species (Feldhamer et al. 2007: table 11.2). Of the seven orders, all contain mammals weighing less than 5 kg. Marsupials (metatherians) are often characterized by the female's abdominal pouch, or *marsupium,* which gives rise to the common name of this group. This is a poor diagnostic feature, however, because not all marsupials have a marsupium, and a pouch occurs in echidnas. A pouch probably is a derived condition in marsupials; only about 50% of species have a permanent pouch. Marsupials are best distinguished from eutherians on the basis of their reproductive mode—specifically, the relatively small maternal energy investment in young before birth. In fact, no marsupials have litters that weigh more than 1% of the mother's body mass (Russell 1982). In contrast, small eutherians, such as rodents or "insectivores," may have litters that weigh 50% of the mother's body mass. Maternal investment in **lactation** is much greater in marsupials, however, so by the time young are weaned, total investment in a litter may be similar in marsupials and eutherians of similar body weight. In addition to reproductive characteristics, marsupials differ from eutherians in many skeletal and anatomical features (Feldhamer et al. 2007: table 11.1).

As noted by Lee and Cockburn (1985), several life history characteristics of marsupials differ from those of eutherians, at least in degree. Different characteristics should not be viewed as "shortcomings," however, but simply as different adaptive strategies. In each case, marsupials show less diversity in adaptive radiation, form, and function than eutherians. For example, marsupials generally have lower basal **metabolic rates** (energy expenditure, measured in kilojoules per day)—about 70% the rates of comparably sized eutherians. Except for bandicoots, marsupials also have slower postnatal growth. Relative brain size also is smaller in marsupials, especially in large species, as is the range of body size. For example, the difference in body mass between the smallest living marsupial species, the Pilbara ningaui (*Ningaui timealeyi,* average body mass 2–9 g, or 0.07–0.31 oz), and the largest, the red kangaroo (*Macropus rufus,* average mass 66 kg, or 145 lb), is about four orders of magnitude. This is much less pronounced than the extremes noted in eutherians—that is, the seven orders of magnitude size difference between pygmy shrews and blue whales. Nor have marsupials developed true flight, like the bats. Consequently, marsupials have been unable to take advantage of the abundant feeding niches afforded by night-flying insects. Likewise, there are no **fossorial** (digging or burrowing) herbivores among the marsupials, and all the large marsupial carnivores are extinct. Marsupials are therefore more restricted than eutherians in their adaptive radiation and associated structural diversity. Nonetheless, a fascinating array of behavioral and morphological adaptations is evident among the seven orders and approximately 330 extant species of marsupials.

Afrosoricida, Erinaceomorpha, Soricomorpha, Macroscelidea, Scandentia, and Dermoptera

Six diverse orders, the Afrosoricida, Erinaceomorpha, Soricomorpha, Macroscelidea, Scandentia, and Dermoptera, represent groups rich in diversity of form and function. Within these orders, 10 extant families comprise about 529 species, representing about 9.7% of all species of mammals. Of the 10 families, all contain mammals weighing less than 5 kg. This rather primitive mammalian assemblage includes: (1) Afrosoricida: tenrecs and otter shrews (family Tenrecidae) and golden moles (Chrysochloridae); (2) Erinaceomorpha: hedgehogs, moonrats, and gymnures (Erinaceidae); (3) Soricomorpha: solenodons (Solenodontidae), shrews (Soricidae), and moles, desmans, and shrew-moles (Talpidae); (4) Macroscelidea: sengis or elephant shrews (Macroscelididae); (5) Scandentia: treeshrews and pen-tailed treeshrews (Tupaiidae and Ptilocercidae); and (6) Dermoptera: colugos or flying lemurs (Cynocephalidae). The Afrosoricida, Erinaceomorpha, and Soricomorpha (formerly in order Insectivora, but as a result of recent molecular research, the taxonomy has changed; in this book I use "insectivores" to refer to members of these three orders) include families that demonstrate **convergence** (the evolution of similar adaptations in distantly related lineages as adaptive solutions to similar ecological pressures)—the talpids and chrysochlorids. The "insectivores" retain many primitive eutherian characteristics, including their brain anatomy, dentition, cranial morphology, postcranial structures, and **cloaca** (a common opening for the urinary and reproductive tracts). As such, they are thought to be ancestral to other mammalian orders. In contrast to their primitive characteristics, many "insectivores" have developed highly specialized features, such as **echolocation** (tenrecs and shrews) and toxin in the saliva (solenodons and some shrews). Unlike many other orders, no key morphological character serves to identify the Afrosoricida, Erinaceomorpha, and Soricomorpha. Elephant shrews (sengis), treeshrews, and colugos were also formerly included in the order Insectivora. Each group is now placed in its own order: the Macroscelidea, Scandentia, and Dermoptera, respectively. The taxonomic history of these orders has been a mystery since the early 1800s and is still being resolved today.

Chiroptera

Second only to rodents in the number of species in an order, bats represent unparalleled variety among the mammals. Bats comprise some 18 families, 197 genera, and more than 1,110 species—about 21.8% of all species of mammals. All 18 families in the order Chiroptera contain mammals weighing less than 5 kg. No order is more diverse than chiropterans in the number of feeding niches they fill. Most bats are **insectivorous**; numerous other species are specialized to feed on fruit, nectar, or pollen. Other bats are **carnivorous**, taking small **terrestrial** vertebrate prey, including other bats, and a few species feed on small fish or a diet of blood. Bats also exhibit diversity in their reproductive patterns and roosting ecology. Although many species have spontaneous ovulation, vespertilionids in northern areas employ delayed fertilization in conjunction with **hibernation** or migration. Other species have **delayed implantation** or delayed development (see Chapter 11 on these reproductive strategies).

Roosting ecology likewise varies among chiropteran species. Bats may roost in large colonies, in smaller aggregations, or in clusters of a few individuals, or they may be solitary. Morphological diversity also characterizes the two suborders: the megachiropterans and the microchiropterans. Most bats are small—many with body mass less than 10 g (0.35 oz). The smallest known mammal on earth is a microchiropteran, the hog-nosed bat

(*Craseonycteris thonglongyai*) of Southeast Asia (family Craseonycteridae), weighing 1.5 to 2.0 g (0.05–0.07 oz). The largest bats are the megachiropterans, family Pteropodidae, which may reach 1.2 kg (2.6 lb) and have wingspans close to 2 m (6.5 ft). Size is only one of the differences between the suborders. Compared with microchiropterans, the facial characteristics of megachiropterans are simple (Feldhamer et al. 2007: table 13.1). They have no nose leaf or other ornamentation, the eyes are relatively large, and the ears (**pinnae**) are without a **tragus** (a projection from the lower margin of the ear). Several other differences exist between the two suborders. Microchiropterans have a sophisticated system of acoustic orientation called **echolocation**. They sense their immediate environment and "locate" food or objects from the echoes returned from high-frequency sound pulses generated and emitted through either the mouth or nostrils. Despite many similarities in the general aspects of echolocation in bats, species differ in the characteristics of sound pulse duration, timing, frequency modulation, intensity, and length of signals. Echolocation does not occur in megachiropterans. These differences between the two suborders have led some authorities to suggest that the pteropodids were derived independently through evolutionary time (i.e., they are diphyletic). Recent genetic evidence suggests, however, that bats evolved from a common ancestor (they are monophyletic), although relationships among families remain unresolved.

Despite the diversity found throughout the order, there is an obvious unifying characteristic: all bats have highly modified forearms and hands that form wings, and they are the only mammals that fly. Wings are the primary structures used to create the upward lift and forward thrust necessary for flight. Although there are common elements in wing structure for flight, there is diversity in the shape of wings. Wing shape and size, quantified in measures of "aspect ratio" and "wing loading capacity" (see Chapter 2), reflect the habitat, foraging characteristics, degree of maneuverability, and general life history characteristics of bat species.

Primates

The order Primates, which includes humans, comprises 15 families, 69 genera, and 376 species; considerable revision of this order has occurred during the past several years, resulting in more families, genera, and species. The order Primates includes about 8% of all species of mammals. Of the 15 families in the order, 13 contain mammals weighing less than 5 kg. Primates are largely tropical and subtropical in distribution, ranging through Africa, Asia, South and Central America, the Malay Archipelago, Japan, and Madagascar. Major characteristics of the group include refined hands and digits, with nails replacing claws; binocular stereoscopic vision; a complete **postorbital bar** (a bony rod that separates the orbit from the temporal pit); a reduced muzzle; and slower rates of reproduction, with increased developmental time. Primates exhibit a progression of socio-spatial systems ranging from overlapping **home ranges** to a diverse array of social and mating systems. Cheekteeth are **bunodont** and **brachyodont**. Primates, as a group, are **generalists** compared with most other mammalian groups.

The extant primates are divided into the suborders Strepsirhini and Haplorhini. Strepsirhines are characterized by a rhinarium (an area of moist, hairless skin surrounding the nostrils), a **bicornuate uterus, epitheliochorial placentation**, relatively small neonates, and a maximum of 36 teeth. The group includes the families Daubentoniidae (aye-ayes), Lorisidae (lorises), Galagidae (galagos), Lemuridae (lemurs), Lepilemuridae (sportive lemurs), Cheirogaleidae (dwarf and mouse lemurs), and Indriidae (indrid lemurs and sifakas). Although the majority of strepsirhines are concentrated in Madagascar, some groups live in Africa and Asia, including the Malay Archipelago. Many of the Malagasy (Madagascan) species are endangered due to habitat destruction.

Haplorhines are characterized by a fused **simplex uterus; hemochorial placentation**; larger neonates than strepsirhines relative to the mother's size; differences in the rhinarium compared with strepsirhines; and a maximum number of teeth of 32 in anthropoid primates from Asia and Africa (Old World primates) and 36 in those from Central and South America (New World primates). Distinctive differences in the visual and olfactory systems of haplorhines also separate them from strepsirhines (Feldhamer et al. 2007). The group includes the families Tarsiidae (tarsiers), Cebidae (marmosets, tamarins, lion tamarins, squirrel monkeys, and capuchin monkeys), Aotidae (night monkeys), Atelidae (spider monkeys, howler monkeys, woolly monkeys, and muriquis), Pitheciidae (saki monkeys, bearded saki monkeys, uakaris, and titi monkeys), Cercopithecidae (cercopithecine and colobine monkeys), Hylobatidae (gibbons and siamangs), and Hominidae (apes and humans). Nonhuman haplorhines are widely distributed throughout Africa, Asia, the Malay Archipelago, and Latin America.

Carnivora

The order Carnivora comprises some 15 families, 126 genera, and 286 species—about 5.6% of all species of mammals. Of the 15 families in this order, 7 contain mammals weighing less than 5 kg. Carnivores are terrestrial or aquatic predators that usually consume other animals as a major part of their diet. Most morphological and behavioral characteristics of carnivores involve adaptations to enhance locating, capturing, killing, and consuming their prey. Within the order Carnivora, some specialized feeding habits have evolved. These include insectivory, as in aardwolves and the sloth bears; scavenging on carrion in hyenas; **omnivory** in many species; and **herbivory** in the greater and lesser pandas. All carnivores have digits with well-developed claws and dentition with enlarged **canine** teeth. The defining ordinal characteristic, however, is **carnassial** teeth—specialization of the fourth upper **premolar** (P4) and first lower **molar** (m1) for cutting and shearing (Elbroch 2006: 47). As described in Chapter 4, carnassial dentition is especially well developed in highly predaceous families, such as felids, canids, mustelids, and hyaenids, and less developed in more omnivorous groups, such as ursids and procyonids. Facial musculature is well developed in the carnivores. The articulation of the jaw with the cranium typically is hinged so as to reduce lateral motion as captured prey struggle to escape. A great deal of variation in body mass occurs among members of this order. The largest carnivores are the southern elephant seals—males are reported to weigh up to 2,200 kg (4,850 lb)! The largest terrestrial carnivores are the polar bear and grizzly bear, which are 11,000 times heavier than the smallest carnivore, the least weasel. For the most part, secondary sexual dimorphism is evident; males are often many times larger and heavier than females. Although most species are solitary hunters, wolves, spotted hyenas, lions, and some other carnivores generally hunt in packs. Thus they are able to prey on species that are several times larger than themselves. In addition to improved foraging efficiency, group membership may provide several additional benefits to individuals: communal infant care, reduced predation, and defense of feeding areas from rival packs.

Two suborders of modern carnivores are generally recognized. Six families occur in the suborder Feliformia: cats (Felidae), mongooses (Herpestidae), hyenas (Hyaenidae), civets and genets (Viverridae), Madagascar mongooses (Eupleridae), and the African palm civets (Nandiniidae). The suborder Caniformia includes nine families: dogs (Canidae); bears (Ursidae); weasels, otters, and badgers (Mustelidae); raccoons, coatis, and allies (Procyonidae); red pandas (Ailuridae); skunks (Mephitidae); and three families of aquatic carnivores (the pinnipeds)—eared seals (Otariidae), earless seals (Phocidae), and the walruses (Odobenidae). Pinnipeds have distinctive morphological adaptations consistent with their aquatic life histories, and in the past they were sometimes placed in their own order.

Rodentia

Rodents constitute the largest order of mammals, with 32 extant families and approximately 2,277 species (Carleton and Musser 2005). Rodents represent about 42% of all living species of mammals in the world today. Except for the family Castoridae (beavers), all families of rodents contain members that weigh less than 5 kg. Capybaras (the largest rodent) were formerly in their own family (Hydrochoeridae) but are now included in the family Caviidae (guinea pigs, cavies, Patagonian "hares," and capybaras). Rodents exhibit a cosmopolitan (worldwide) distribution and are native everywhere except Antarctica, New Zealand, and a few oceanic islands. They have adapted very successfully to a wide range of terrestrial, arboreal, **scansorial** (adapted for climbing), fossorial, and semiaquatic habitats. Rodents are found in all biomes, often as **commensal** (living in close association) with humans. They have a diverse array of locomotor adaptations, including **plantigrade** (walking on the soles of the feet), **cursorial**, swimming, fossorial, jumping, and gliding. The majority of rodents are small (20–100 g, or 0.7–3.5 oz), although the largest, the capybara (*Hydrochaerus hydrochaeris*), may reach 50 kg (110 lb). Given the large number of rodent species, their degree of diversity and adaptability, and their convergent evolutionary trends, it is not surprising that the systematic relationships of many families and subfamilies are complex and result in an array of suborders, superfamilies, and subfamilies.

Despite the number of species and their widespread distribution and diversity, rodents are surprisingly uniform in several general morphological characteristics. The diagnostic characteristic that defines all rodents is a single pair of upper and lower **incisors** (Elbroch 2006: 264). Their large incisors are open-rooted, ever-growing, and used for gnawing (the name "Rodentia" is derived from the Latin *rodere*, "to gnaw"). In many species, the mouth can close behind the incisors, and the animal may have either internal or external cheek pouches for transporting food. Rodents also have a **diastema** (a gap between the incisors and cheekteeth) that allows for maximum use of the incisors in manipulating food. Canine teeth are absent, and the number of molariform teeth is reduced. Molariform dentition may or may not be ever-growing, and many different **occlusal cusp** patterns occur. The typical total number of teeth for rodents is 16. Rodents never have more than two pairs of premolars, and only one species has more than 22 teeth—the silvery mole-rat (*Heliophobius argenteocinereus*), with 28. Rodents generally are herbivorous or omnivorous, and a few are heavily insectivorous, depending on the season and availability of food items. **Coprophagy** (**reingestion** of their own fecal pellets, taken directly from the anus) has been reported in 11 families of rodents (Hirakawa 2001). Females have a **duplex uterus**, males have a **baculum** (penile bone), and the testes may be scrotal only during the breeding season.

Despite the confusion and lack of consensus in rodent classification, three groupings generally are accepted and useful for defining the groups of rodents: **sciuromorph** (squirrel-shaped), **myomorph** (mouse-shaped), and **hystricomorph** (porcupine- or cavy-shaped) rodents (Elbroch 2006: 262; Feldhamer et al. 2007: fig. 18.3). Each group is distinguished by skull structure and jaw musculature—specifically, the origin of the enlarged chewing muscle, the **masseter**, which permits both vertical and forward-and-backward motion of the lower jaw. These groups are often given subordinal status, although not all families readily fit into just one of the groups.

The sciuromorph rodents have a deep and short masseter muscle extending from the mandible to the **zygomatic arch**. Only a very small opening (**infraorbital foramen**) is present on the front of the zygomatic arch, through which no masseter passes. In this condition, the upper part of the masseter reaches the back of the skull, with the deep part extending to the zygomatic arch; the **temporalis** muscle is small. This arrangement permits a strong forward motion when biting. Representative families include beavers (Castoridae), mountain beavers (Aplodontidae), squirrels (Sciuridae), pocket gophers (Geomyidae), scaly-tailed squirrels (Anomaluridae), pocket mice (Heteromyidae), and springhares (Pedetidae). Squirrel-like rodents have four to five cheekteeth in each row, compared with three in the mouselike (myomorph) rodents. In comparison with other rodents, sciuromorph rodents have rather simple teeth that lack strong projecting cusps or sharp ridges of enamel on the chewing (occlusal) surfaces. The teeth of beavers, however, depart from this trend by having a well-defined pattern of ridges that are adaptive in their diet of bark and fibrous vegetation.

Squirrels (family Sciuridae) reside throughout the world except for Australia, Polynesia, Madagascar, southern South America, and some desert regions. Most are seed-eaters, and they are the world's principal arboreal herbivores. Most squirrels are instantly recognizable by their cylindrical bodies, long whiskers, large eyes, bushy tails, and pronounced manual dexterity; ground squirrels have shorter and less bushy tails than the tree squirrels. Most squirrels are familiar to humans because of their **diurnal** activity and tendency to stockpile their groceries in **cache** sites in almost every nook and cranny. We think of squirrels as primarily arboreal; however, there are plenty that do not reside in trees—namely, chipmunks, ground squirrels, prairie dogs, and marmots. One of the largest members of the squirrel family common in eastern North America is the woodchuck (a.k.a. "Punxsutawney Phil")—a rather substandard weather forecaster from central Pennsylvania! The semiaquatic beavers (*Castor canadensis*) of the family Castoridae, equipped with webbed feet, do not look much like squirrels; however, their "zygomasseteric complex" is squirrel-like. The mountain beaver (*Aplodonta rufa*) of the Pacific Northwest of North America, the most primitive of all present-day rodents, is also included in this group. Sciuromorph rodents include the well-known true hibernators such as Richardson's, arctic, golden mantled, and thirteen-lined ground squirrels, marmots, and chipmunks, to mention just a few.

In myomorph rodents, both the lateral and deep masseter muscles extend far forward, with the deep masseter passing from the lower jaw through the eye socket (orbit) and infraorbital foramen and attaching on the muzzle. This arrangement provides a very effective gnawing action. The temporalis muscle is large and permits a versatile chewing action. Representative families include Madagascar rats and mice (Nesomyidae); mouselike hamsters (Calomyscidae); voles and mice (Cricetidae); Old World rats and mice (Muridae); spiny and soft-furred tree mice (Platacanthomyidae); dormice (Gliridae); jerboas, birch mice, and jumping mice (Dipodidae); and zokors, bamboo rats, and blind mole-rats (Spalacidae). Most mouselike rodents have three cheekteeth in each row. Hamsters have low-crowned teeth with rounded cusps, whereas the rats and mice have a more complex biting surface. Voles and lemmings have a complex occlusal surface of ridges of hard enamel, adaptive in masticating abrasive plant walls of grasses, a staple in their diet. Mice are found in almost all the world's terrestrial habitats except Antarctica. More than one-quarter of all species of mammals are rats and mice. They are typically small with a pointed face and long whiskers. Most are nocturnal and eat seeds.

Ecologically, many of the mouselike rodents may be classified as r-strategists (**r-selected** species)—they are adapted to climatically stressful environments such as montane and arctic ecosystems. They have high reproductive rates, young develop rapidly (they are **precocial**), their body mass is small, reproduction is prolific, life spans are comparatively short, and adults provide minimal parental care to the young. A good example of an r-selected species is the southern red-backed vole *Myodes* (formerly *Clethrionomys*) *gapperi,* an inhabitant of montane forests of North America. This contrasts with **K-selected** species, which reside in more stable environments such as tropical regions, have **altricial** young that require significant parental care, and have low intrinsic rates of increase, low fecundity, and later maturation. A good example of a K-selected species is the California mouse (*Peromyscus californicus*), an inhabitant of the chaparral of northern California. Many species, however, are intermediate between these extremes (Feldhamer et al. 2007). I like to use these two species—*M. gapperi* and *P. californicus*—as examples because they are familiar to me from my graduate school research in California and Colorado. There are many good examples of species that conform to this r-K continuum. Many researchers reject the concepts of r- and K-selection (Stearns 1992); however, many mammalogists find it a useful way to classify and compare life history patterns of mammals.

Mouse-shaped rodents are touted as the most important small mammals, both in terms of their effect on the environment and as a staple food item for many predators. This group is home to such universal pest species as Norway rats, roof rats, and house mice and includes the voles and lemmings—known from popular legend for their suicidal marches, en masse, to the sea (but, for a reality check, see Chapter 13).

In hystricomorph rodents, the deep part of the masseter muscle extends through a large opening in the front of the zygomatic arch and attaches on the muzzle just in front of the eye. This positioning provides the very powerful gnawing action so characteristic of this group of rodents. The lateral masseter is only used in closing the jaw, and the temporalis muscle is small. Most porcupine- or cavy-shaped rodents have robust, angular skulls equipped with large, strongly developed incisor teeth. Their cheekteeth are four in number and show a great variation in occlusal patterns that tracks the diversity of diets, ranging from the aquatic grasses diet of the capybara to the roots and tubers diet of the fossorial tuco-tucos. Representative families include New World porcupines (Erethizonidae); Old World porcupines

(Hystricidae); mole-rats (Bathyergidae); viscachas and chinchillas (Chinchill-idae); pacaranas (Dinomyidae); cavies, Patagonian "hares," guinea pigs, and capybaras (Caviidae); gundis (Ctenodactylidae); dassie rats (Petromyidae); cane rats (Thryonomyidae); agoutis and acouchis (Dasyproctidae); pacas (Cuniculidae); tuco-tucos (Ctenomyidae); viscacha rats and coruros (Octo-dontidae); chinchilla rats (Abrocomidae); spiny rats (Echimyidae); hutias (Capromyidae); and nutrias (Myocastoridae). This diverse group of rodents occurs in North and South America and Asia and includes the largest living rodent, the capybara, and the New World porcupine with its arsenal of spines and **prehensile** tail. The most bizarre of the cavy-like rodents may be the mole-rats of eastern Africa. These virtually hairless rodents spend almost their entire life belowground in extensive colonies, showing similarities to social insects. Most species are typified by a relatively large head, plumb body, short tail, and slender legs. Most of the American hystricomorph ro-dents are terrestrial and herbivorous, although some, such as the porcu-pines, are arboreal, and members of one group (tuco-tucos) are burrowers. In general the group is characterized by production of small litters after a long **gestation period** that results in well-developed (precocial) young. For instance, guinea pigs typically produce two to three young following a ges-tation period of 50 to 75 days, compared with seven to eight young after only 21 days for the Norway rat (*Rattus norvegicus*).

Lagomorpha

The order Lagomorpha (meaning "hare-shaped') includes 11 genera and about 54 species of rabbits and hares in the family Leporidae and 1 genus (*Ochotona*) and about 30 species of pikas in the family Ochotonidae. All members of this order weigh less than 5 kg. Lagomorphs occur worldwide except for the southern portions of South America, Australia, and New Zealand, and islands such as Madagascar, the Philippines, and those in the Caribbean.

All lagomorphs are small to medium-sized terrestrial herbivores. Co-prophagy is common in lagomorphs, as it is in many rodents. Leporids reingest soft fecal pellets, which enhances their ability to live on relatively low-quality vegetation. The diagnostic feature of the order is the occurrence of peg teeth, a second pair of small incisors without a cutting edge, imme-diately behind the larger, rodentlike first incisors (Elbroch 2006: 248). As in rodents, canines are absent and a large space (diastema) separates the incisors and the first cheektooth. The cheekteeth are rootless and **hypsodont**. The crowns of the cheekteeth are relatively simple, with transverse basins separated by enamel ridges. The total number of teeth in lagomorphs ranges from 26 to 28. The cutting edge of the primary incisors is notched in pikas but not in rabbits and hares. The cheekteeth and incisors are open-rooted and ever-growing. In leporids, but not ochotonids, the **rostral** portion of the maxilla is fenestrated (having small, latticelike perforations in the bone) and the frontal bone has a supraorbital process. Rabbits and hares generally have large ears and elongated hind limbs to accommodate their **saltatorial** (leaping) locomotion. Rabbits have a familiar "cotton-ball" tail, but the tail in hares is longer. Pikas are rodentlike in appearance and have short limbs, small ears, and no tail. Although agile, they do not exhibit the running ability of leporids. Unlike rodent feet, the feet of lagomorphs are fully furred. A cloaca is present, the uterus is duplex, and there is no baculum.

Hyracoidea

The order Hyracoidea includes 11 species in three genera, all in the family Procaviidae. This small order comprises about 0.18% of all species of mammals. Referred to in the Bible as "rock badgers" and "conies," the hyraxes are members of a single family, Procaviidae, with three genera and four species. All members weigh less than 5 kg. The family name means "before the caviids (guinea pigs)" and indicates the taxonomic confusion that has surrounded hyraxes. Because of their superficial resemblance to rodents, hyraxes were initially grouped with guinea pigs. The confusion never ends—even the common name "hyrax," meaning "shrew mouse," is misleading. Neither of these associations is accurate; based on fossil and molecular evidence, hyracoids are most closely related to elephants and manatees!

Hyraxes live in central and southern Africa, Algeria, Libya, Egypt, and parts of the Middle East, including Israel, Syria, and southern Saudi Arabia. The rock hyrax (*Procavia capensis*) is the most widely distributed geographically and elevationally. It inhabits rocky outcrops from sea level to 4,200 m (13,800 ft) elevation in Africa and the Middle East. The yellow-spotted rock hyrax (*Heterohyrax brucei*) is found in similar rocky habitats in northeast to southern Africa. The two species of arboreal tree hyraxes (*Dendrohyrax*) inhabit forested areas of Africa up to 3,600 m (11,800 ft) elevation. The terrestrial species are diurnal or crepuscular and form large colonies. Conversely, tree hyraxes are nocturnal and solitary. Hyraxes are short, with compact bodies and very short tails. **Pelage** color is brown, gray, or brownish yellow. Total length varies from 32 to 60 cm (12.5–23.5 inches) and body mass ranges from 1 to 5 kg (2.2–11 lb). *P. capensis* is larger than *H. brucei*. There is no sexual dimorphism. In Chapter 3, I discuss the natural history of "rock badgers" and how they cope with their high-fiber diet.

PART ONE

Modes of Feeding

....

Small mammals, like all organisms, require energy and nutrients for maintenance, growth, activity, and reproduction. Maintaining a high body temperature, which is a key feature of the class Mammalia, also requires regular acquisition of food. The food of small mammals ranges from highly mobile insects, a staple in the diet of shrews and bats, to sedentary forms such as plants, used by the most abundant mammals—the rodents. Small mammals consume food of high energy content (such as blood of vertebrates and insects) as well as food of low energy value (grasses and stems). Earthworms are a staple in the diet of many shrews and moles. Huge numbers of rodents are mainly seed-eaters, with a few (such as the grasshopper mouse) that feed on insects, slugs, snails, spiders, centipedes, and millipedes. The food may be highly specialized and restricted (such as the nectar of localized plants) or rather general and readily available (grasses and herbs). To meet their high energy needs, small mammals have evolved a diverse suite of adaptive responses, as depicted in their morphology, physiology, and behavior. The radiation of food-gathering adaptations is diverse and reflects the tremendous breadth of available food types.

The feeding apparatus of mammals paints a picture of the foods they consume. Adaptations for capturing food (teeth, tongue, and jaw musculature) and processing it (alimentary canal) are crucial to understanding the ecology of a species; feeding also integrates the sense organs and locomotor adaptations. Food habits cannot be used as a systematic criterion for classification, because many members of a given order may not fit neatly into the category defined by the ordinal name. Some examples are rodents such as grasshopper mice, which feed primarily on insects; northern flying squirrels (*Glaucomys sabrinus*) and jumping mice (*Zapus* and *Napaeozapus*), which feed on subterranean fungi (Endogonacea); and vampire bats (*Desmodus*), which feed on blood.

All mammals, except certain whales, monotremes, and anteaters, have teeth, and these structures are inextricably linked with food habits. As mammals evolved in the Mesozoic Era, major changes occurred in their dentition and jaw musculature; teeth became differentiated to perform specialized functions. Within extant (living) species, several trophic groups can be recognized: insectivorous, herbivorous, carnivorous, and omnivorous mammals. Other specialized modes of feeding have evolved from these four basic plans (see Feldhamer et al. 2007: fig. 7.1).

TWO

Insectivory

Insects are ubiquitous, occurring in the air, on plants, in soil, and in water. Mammals that consume insects, other small arthropods, or worms are referred to as **insectivorous** ("insect-eating"). Examination of mammalian fossils from the Triassic Period has revealed that the insectivorous feeding niche represented the primitive, or basal, condition of eutherian (**placental**) mammals. Insects are highly digestible and provide a high yield of metabolizable energy per unit intake; as a result, this dietary source is very desirable for small mammals. Today, this feeding niche is exploited by members of 13 mammalian orders: echidnas and platypuses (Monotremata); hedgehogs, shrews, and moles (Notoryctemorphia, Afrosoricida, Erinaceomorpha, and Soricomorpha); most bats (Chiroptera); anteaters and armadillos (Cingulata and Pilosa); the pangolins (Pholidota); aardvarks (Tubulidentata); and the aardwolf (Carnivora). The noteworthy diminutive insect-eaters include the shrew opossums (order Paucituberculata, family Caenolestidae), found in South America; the nearly cosmopolitan (worldwide) shrews and moles; North American rodents such as the grasshopper mice; and finally, the dunnarts, planigales, and ningauis (order Dasyuromorphia, family Dasyuridae) of Australia and New Guinea.

General Characteristics

Dilambdodont Dentition

To masticate insects, small mammals must be able to crush and dissect tough, chitinous exoskeletons. The dentition of hedgehogs, shrews, moles, and insectivorous bats is typified by a continuous row of **molars** equipped with needle-sharp **cusps** (Elbroch 2006). For crushing exoskeletons, molar

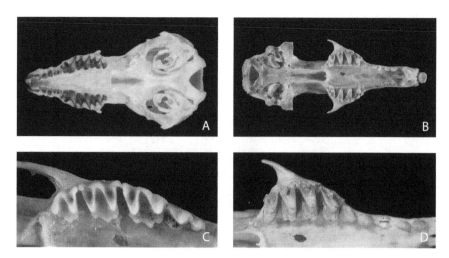

FIGURE 2.1. Ventral views of the cranium of (A) a northern short-tailed shrew (*Blarina brevicauda*) and (B) a Haitian solenodon. Occlusal views of (C) dilambdodont teeth (cusps form a W shape), found in shrews, moles, and many insectivorous bats, and (D) zalambdodont teeth (cusps form a V shape), found in golden moles, solenodons, tenrecs, and otter shrews. *Source:* University of Michigan, Animal Diversity Web, http://animaldiversity.org, courtesy of Phillip Myers.

teeth have a chewing (**occlusal**) surface characterized by many sharp ridges that form a W-shaped pattern, referred to as **dilambdodont**. Examples of mammals with dilambdodont teeth include shrews, moles, and many insectivorous bats (e.g., Vespertilionidae). Molar teeth characterized by occlusal surfaces with V-shaped crests are referred to as **zalambdodont**; this dentition is found, for example, in golden moles, solenodons, tenrecs, and otter shrews (fig. 2.1).

Shrews and moles have lower **incisors** that are slightly **procumbent** (pointing forward and upward) to aid in grasping prey (fig. 2.2), acting like small forceps or pincers. In members of the subfamily Soricinae, the red-toothed shrews, the tips of the teeth are a deep red. This color is caused by iron deposits and is worn down and reduced as individuals age. Shrews with pigmented teeth are thus distinguished from the white-toothed shrews (subfamily Crocidurinae), which do not have this characteristic. The two groups also differ in biogeography, behavior, and physiology (Feldhamer et al. 2007). Although shrews consume primarily invertebrates, flexibility of food habits ensures survivorship, especially in harsh environments. For

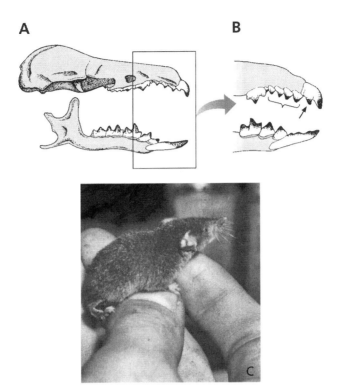

FIGURE 2.2. Characteristic features of shrew skulls. (A) A typical shrew skull. Note the lack of auditory bullae and zygomatic arch. (B) The enlarged anterior portion of the skull of a red-toothed shrew, showing the distinctively pigmented enamel, procumbent first lower incisor, secondary cusp on the first upper incisor (arrow), and unicuspid teeth (bracket). The incisors function as tweezers, picking up insects that are then passed to the sharp, multicuspid posterior teeth for crushing and chewing. (C) A lesser white-toothed shrew (*Crocidura suaveolens*). *Sources:* (A, B) Feldhamer at al. 2007; (C) Werner Haberl.

example, in coniferous forests of northeastern Siberia, larch (*Larix*) seeds were found to be the staple in the winter diet of Laxmann's shrews (*Sorex caecutiens*) (Dokuchaev 1989). **Terrestrial** shrews such as greater white-toothed shrews (*Crocidura russula*) and northern short-tailed shrews (*Blarina brevicauda*) capture lizards and small mammals. The semiaquatic species such as Eurasian water shrews (*Neomys fodiens*) and American water shrew (*Sorex palustris*) are avid consumers of frogs, newts, and even small fish.

Terrestrial Insectivores

In the class Mammalia, terrestrial insects form the staple in the diet of a diverse group of small mammals with unique specializations adaptive for eating insects. There is a great deal of evolutionary convergence in form and function among these groups. However, as noted in Chapter 1, the taxonomy of order Insectivora has changed and I use "insectivores" to refer to members of three orders of mammals: Afrosoricida, Erinaceomorpha, and Soricomorpha (Wilson and Reeder 2005). Six diverse families comprise this rather primitive mammalian assemblage: hedgehogs, moonrats, and gymnures (Erinaceidae); tenrecs and otter shrews (Tenrecidae); golden moles (Chrysochloridae); shrews (Soricidae); moles, desmans, and shrew-moles (Talpidae); and solenodons (Solenodontidae). Because insectivorous mammals consume minimal amounts of fibrous vegetative material, prolonged intestinal fermentation is not required; their alimentary canals are short, and most "insectivores" and chiropterans lack a cecum.

Shrews constitute the largest and most widely distributed family of "insectivores." There are 26 genera and 376 species, although many are of uncertain taxonomic status (Hutterer 2005). This large family (7% of all species of mammals) is divided into two subfamilies. The red-toothed shrews (subfamily Soricinae) include three tribes found throughout much of the Nearctic, Palaearctic, and Oriental faunal regions (fig. 2.3). The white-toothed shrews (subfamily Crocidurinae) are restricted to Old World faunal regions. Shrews are generally small, with diminutive eyes and a long, pointed rostrum (snout). Their pelage is short, dense, and usually dark-colored. In many species, lateral glands produce a musky odor that is most noticeable during the breeding season. The adult Etruscan shrew (*Suncus etruscus*)—arguably, the smallest terrestrial mammal in the world—has a body mass of 2 g (0.07 oz) and a combined head and body length of 35 mm (1.4 inches). The American pygmy shrew (*Sorex hoyi*) of North America, averaging 2 g in body mass, is also touted as the smallest terrestrial mammal. At the other end of the spectrum is the Asian house shrew (*Suncus murinus*), at 100 g (3.5 oz), with a head and body length of 150 mm (5.8 inches). Most shrew species weigh 10 to 15 g (0.35–0.52 oz) and have a head and body length of about 50 mm (1.9 inches).

Shrews are opportunistic predators, devouring a wide array of invertebrates, primarily beetles, bugs, earthworms, woodlice, spiders, snails, and

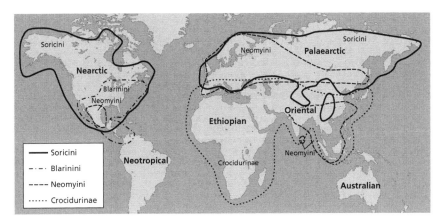

FIGURE 2.3. Geographic distribution of red-toothed shrews (subfamily Soricinae, which includes three tribes: Neomyini, Soricini, and Blarinini) and white-toothed shrews (subfamily Crocidurinae). *Source:* Feldhamer et al. 2007, data from S. Churchfield, *The Natural History of Shrews* (Cornell Univ. Press, 1990).

insect larvae (Churchfield 1990). As with other small mammals, the diet of shrews tends to coincide with availability of insects. However, shrews do show preferences. The common shrew (*Sorex araneus*) is reported to select certain species of terrestrial woodlice over others. Millipedes, a common inhabitant of forest litter, are rarely consumed by shrews, while centipedes often are. Snails such as *Oxychilus alliarius*, like millipedes, may be avoided because of their acrid secretions. Earthworms are a favored food of many species of shrews, yet are rarely eaten by Eurasian pygmy shrews (*Sorex minutus*); this may be attributable to earthworms being too large for the tiny pygmy shrews to handle.

Venomous Saliva

Four species of mammals produce a venomous saliva (fig. 2.4): the northern short-tailed shrew (*B. brevicauda*) of North America, the Eurasian water shrew (*Neomys fodiens*), the Mediterranean water shrew (*N. anomalous*), and the Haitian solenodon (*Solenodon paradoxus*). Unfortunately, the extant species of solenodon is endangered and narrowly distributed on Cuba and Hispaniola. The northern short-tailed shrew, by contrast, is common and widely distributed in many diverse ecological communities of eastern North America. In both *Blarina* and *Neomys*, the toxin is stored in

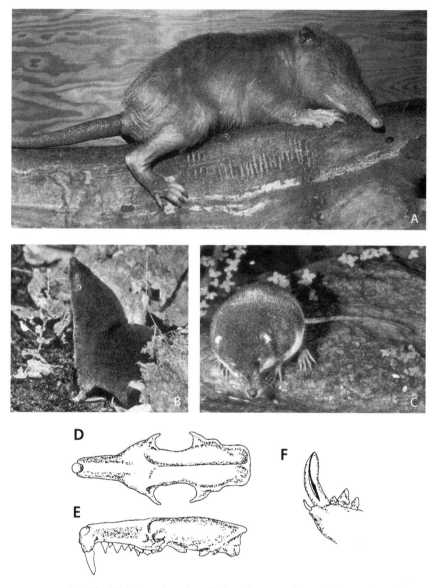

FIGURE 2.4. (A) The Haitian solenodon (*Solenodon paradoxus*), (B) northern short-tailed shrew (*Blarina brevicauda*), and (C) European water shrew (*Neomys fodiens*) are venomous mammals. (D) Dorsal and (E) lateral views of the skull of a Haitian solenodon. Note the incomplete zygomatic arch and enlarged first upper incisor. (F) The large second lower incisor is deeply grooved to accommodate secretions of toxin from the submaxillary gland below it. *Sources:* (A) Dr. Pat Morris / ardea.com; (B) Peter Vogel; (C) Hal S. Korber; (D–F) Feldhamer et al. 2007, adapted from A. F. DeBlase and R. E. Martin, *A Manual of Mammalogy* (2d ed., Wm. C. Brown, 1981).

submaxillary glands and is administered to the prey through a concave medial surface in the first lower incisors (Kido and Uemura 2004). The toxin of *Blarina* was recently purified by Kita and associates (2005) and characterized as a lethal mammalian venom with a tissue kallikrein-like protease (protein-digesting) activity, derived from the submaxillary and sublingual salivary glands. This venom acts mainly on the nervous system, paralyzing tiny victims. The venomous saliva of solenodons and shrews may have evolved to enable these animals to capture larger prey than their body size would otherwise permit. Extracts of the toxin administered to mice affect the nervous, respiratory, and vascular systems, causing irregular respiration, paralysis, and convulsions, followed by death. The ability to cache uneaten prey ensures that a predictable energy source is readily accessible and available when prey is scarce. Research by Tom Tomasi (1978) has demonstrated the importance of venom in the hoarding behavior of *Blarina*. The shrew bites its prey, which is then immobilized. Next, the shrew caches the comatose prey belowground. Caching sites can be numerous, marked by defecation and urination, and provide the shrews with a source of fresh food for some time. (Food hoarding, or **caching**, is discussed in more detail in Chapter 3.)

Northern short-tailed shrews are notorious for their belligerent personality. The well-known naturalist and conservationist Teddy Roosevelt, who also served as twenty-sixth president of the United States, was impressed with the aggressive demeanor of his pet shrew (*B. brevicauda*). He wrote: "certainly a more bloodthirsty animal of its size I never saw." The venomous bite of shrews also gained Hollywood fame in the 1959 B movie *The Killer Shrews*, in which giant venomous shrews attacked and killed humans.

Although the bite of *B. brevicauda* has certainly not been known to kill a human, some adverse effects have been reported. Following the receipt of "four small punctures" near the base of his second finger, nineteenth-century naturalist Charles Maynard carefully documented the events that followed. He noted that in approximately 30 seconds he began to experience a burning sensation:

> The burning sensation, first observed, predominated in the immediate
> vicinity of the wounds, but was now greatly intensified, accompanied
> by shooting pains, radiating out in all directions from the punctures but
> more especially running along the arm, and in half an hour, they had
> reached as high as the elbow. All this time, the parts in the immediate

vicinity of the wounds, were swelling, and around the punctures the flesh had become whitish . . . I bathed the wounds in alcohol and in a kind of liniment, but with little effect. The pain and swelling reached its maximum development in about an hour, but I could not use my left hand without suffering great pain for three days, nor did the swelling abate much before that time. At its greatest development, the swelling on the left hand caused that member to be nearly twice its ordinary thickness at the wound, but appeared to be confined to the immediate vicinity of the bites, and was not as prominent on the right hand; in fact, the first wound given was by far the most severe. The burning sensation disappeared that night, but the shooting pains were felt, with less and less severity, upon exertion of the hand, from the elbow downward, for a week, and did not entirely disappear until the total abatement of the swelling, which occurred in about a fortnight. (Maynard 1889)

When I was teaching the course "Winter Mammalian Ecology" at SUNY College of Environmental Science and Forestry's Adirondack Ecological Center in Newcomb, New York, I witnessed the results of an encounter with *Blarina* by one of my students. While handling a short-tailed shrew with his bare hands, a student was bitten between the thumb and forefinger by a male (reproductively active) shrew. Understandably, the student's first response was to drop the shrew. When he attempted to regrasp the 20 g animal, the shrew responded by inflicting a second bite—at exactly the same site! The shrew was quickly released and ran off into the woods, leaving the student rather embarrassed as his peers and professor cheered the shrew's victory and photodocumented the incident for posterity. At the end of the week, the student sheepishly revealed his right arm to me. His hand exhibited a well-developed area of edema in the vicinity of the wound, accompanied by discoloration of the tissue along his arm, reaching to the elbow. The symptoms and appearance of the "site of injection" were very similar to those described by Maynard (1889). Fellow students were impressed with his battle wounds, but I showed little sympathy, indicating that such wounds are routine and are revered by "real" field biologists!

Long, Protrusile Tongue

Several groups of insectivorous mammals are **myrmecophagous** ("ant-eating"). Representative groups include the armadillos, silky anteaters, giant anteaters, pangolins, aardvarks, and numbats, which feed on colonial insects such as ants and termites (Feldhamer et al. 2007). Most myrmecophagous mammals are not considered "small mammals" and represent a bizarre mix of New World mammals formerly placed in the order Xenarthra and now divided into two separate orders: Pilosa (anteaters and sloths) and Cingulata (armadillos). Reduction in the number of teeth is common among myrmecophagous mammals, and their dentition departs from the "insectivorous" design of the hedgehogs, shrews, and moles. They have numerous peglike teeth (armadillos) or no teeth at all (echidnas, anteaters, and pangolins). Many mammals that consume colonial insects such as termites and ants have long, extensible (protrusile), wormlike tongues (Chan 1995; Reiss 1997). These highly maneuverable, sticky tongues are effective in reaching the inner recesses of ant and termite nests.

Short-beaked echidnas (*Tachyglossus aculeatus*) weigh between 2.5 and 7 kg (5.5–15 lb) and live in a variety of habitats, including forests, sandy plains, rocky areas, and hills of Australia, Tasmania, and New Guinea (fig. 2.5A). The long-beaked echidnas (*Zaglossus attenboroughi, Z. bartoni,* and *Z. bruijni*) range in mass from 5 to 10 kg (11–22 lb) and are restricted to forested highland areas of New Guinea (fig. 2.5B). The diet of *Zaglossus* (meaning "long tongue") differs from that of *Tachyglossus* in that earthworms are the staple. Echidnas locate food using their keen senses of hearing and smell. Both genera of echidnas have a long "beak," which, like the beak of the platypus, contains electroreceptors. They may use these electroreceptors or may sense vibrations to help locate prey. Echidnas also have a long, extensible tongue coated with a sticky secretion produced by enlarged submaxillary salivary glands. The snout is strong enough to use as a tool to break open hollow logs and excavate forest litter to access ants and termites. In addition, each foot is equipped with five digits (**pentadactyl**) with flat claws well adapted for digging. *Zaglossus* ingests earthworms hidden under the leaf litter by capturing them in a groove on its tongue—this is done by extending the tongue 2 to 3 cm (0.8–1.2 inches) beyond the end of the snout and opening the groove by muscle flexion (Griffiths, Wells, and Barrie 1991). Using its sticky tongue, *Tachyglossus* (meaning "rapid tongue") consumes ants, termites, and other insects, which are ground to a paste between the tongue and spiny palatal ridges. Peggy Rismiller and Mike McKelvey, of Kangaroo Island, Australia, found that echidnas on the island, located off the coast of South Australia, feed on a wide variety of invertebrates, including grubs, beetles, nematodes, invertebrate eggs, earth-

FIGURE 2.5. (A) Short-beaked echidna (*Tachyglossus aculeatus*) and (B) western long-beaked echidna (*Zaglossus bruijni*). *Sources:* (A) Courtesy of Peggy G. Rismiller and Michael W. McKelvey (*Journal of Mammalogy* 81:1–17, 2000); (B) D. Parer and E. Parer-Cook / AUSCAPE.

worms, insect larvae, ants, and termites. "Many of the echidnas' food sources are larger than their mouth gape," Rismiller (1999: 99) wrote. "We have observed echidnas placing their mouth opening around soft prey and literally sucking it in. If this fails, it uses its beak as a battering ram, ruptures, the prey and skillfully licks out the nutritious semi-liquid body mass." In August and September, the foraging of echidnas centers on the mounds of meat ants (*Iridomyrmex detectus*), where they feed on fat-laden females. In cool, **mesic** (moderately moist) areas, *Tachyglossus* feeds primarily on ants, but in hot, **xeric** (dry) areas, termites are a staple in its diet. It is unclear why this switch in diet occurs, given the abundance of ants in both environments (M. Griffiths 1968).

Among small mammals, the champion termite-eater is the numbat (banded anteater) of Australia (fig. 2.6). Approximately the size of the eastern chipmunk (*Tamias striatus*) of North America, the numbat is the only small mammal that concentrates on eating termites for a living. It is the sole member of the **marsupial** family Myrmecobiidae, and it has numerous, small, delicate teeth—the total number may be as high as 52. Numbats do eat ants, although only accidentally; the primary, almost exclusive item in their diet is termites. Numbats are the only fully day-active marsupial in Australia; perhaps this is attributable to their diet of termites, a **diurnal** species. Adult numbats are solitary and territorial. They spend most of their waking hours sniffing the ground, flipping small pieces of wood, and busily searching for their termite prey. They have small claws and are unable to

FIGURE 2.6. The numbat (*Myremecobius fasciatus*). *Source:* Bert & Babs Wells / Oxford Scientific.

excavate well-fortified termite mounds. They use their well-developed sense of smell to locate the shallow, unfortified underground galleries that termites construct between the nest and feeding sites.

Once a gallery of termites is located, a numbat feeds by means of its long, protrusile, sticky tongue, which is inserted into the galleries of a nest. Termites that adhere to the tongue are retained in the mouth as the tongue is flicked in and out. Numbats synchronize their foraging day with termite activity, which is temperature-dependent. In winter, numbats feed from midmorning to midafternoon. In summer, they awake early and begin to forage. They then take shelter during the heat of the day and feed again in the late afternoon. At night, numbats retreat to their nest, typically located in a hollow log or tree or in a burrow composed of a narrow shaft 1 or 2 m (3.2–6.5 ft) long, which terminates in a spherical chamber lined with soft plant material such as grasses, leaves, and shredded bark. Numbats were once found across the southern and central parts of Australia. They are now restricted to eucalypt forests and woodlands in the southwest of western Australia. Sadly, they are seriously endangered by habitat destruction for agriculture and predation by introduced foxes and feral house cats.

Arboreal Insectivores

Elongated Digits

Much like the use of a long, protrusile tongue in echidnas and numbats, some **arboreal** primates and marsupials use elongated digits to secure well-hidden insect prey. With its prominent naked ears, long, brushy tail, and elongated third finger, the aye-aye (*Daubentonia madagascariensis*) is distinctive among mammals (Quinn and Wilson 2004) (fig. 2.7A). An inhabitant of moist forests of Madagascar, its appearance is so bizarre that people native to Madagascar have long considered it an omen of bad luck. Weighing in at about 2.2 kg (5 lb), aye-ayes are the largest **nocturnal** primate. The German name for the aye-aye means "finger-beast," pertaining to the particularly long fingers, with the middle finger bearing a long, wirelike claw that is used for extracting insects from wood (Krakauer, Lemelin, and Schmitt 2002). Aye-ayes also use this elongated digit and claw to remove the pulp from fruits such as mangos and coconuts. The aye-aye has chisel-like incisors used for gnawing and chewing, much in the fashion of rodents. Because of the shape of the teeth and the presence of a **diastema** between the incisors and cheekteeth, aye-ayes were first described as rodents.

Using their keen hearing, aye-ayes detect larval insects hidden under the bark of dead branches. They expose prey by first gnawing off the overlying bark with their incisors, then inserting the third finger to crush and extract larvae, which they transfer to their mouth. The insect-eating niche that aye-ayes occupy in Madagascar is exploited by woodpeckers elsewhere in the world (Macdonald 2006).

The higher elevations of forests in New Guinea are home to an example of how similar evolutionary pressures have acted on two different mammals that inhabit two distant islands to produce almost identical feeding specializations. It is a fascinating case of convergent evolution in feeding specializations—those of the aye-aye, a Malagasy primate, and the long-fingered triok (*Dactylopsila palpator*), a New Guinea marsupial (fig. 2.7B).

FIGURE 2.7. (A) The aye-aye (*Daubentonia madagascariensis*) feeds primarily on the tree-burrowing larvae of beetles. It bites into the bark with its powerful incisors and crushes and extracts insect larvae with its elongated third finger. (B) The long-fingered triok (*Dactylopsila palpator*), an inhabitant of New Guinea. *Sources:* (A) Kevin Schafer, Schafer Photography; (B) Peter Schouten.

Like the aye-aye, trioks have a pronounced sense of hearing, adaptive for detecting insect grubs chewing deep within inner recesses of trees and rotting logs. An elongated tongue and sharp, powerful, chisel-like incisors assist in tearing through the bark—entire trees may be scarified by teeth marks. Trioks then use their greatly elongated fourth finger (recall that aye-ayes used their third finger), equipped with a hooked nail, to extract tunneling larvae. As with the North American pileated woodpecker, the feeding activities of trioks can be detected by the presence of woodchips at the base of the excavation site. A rather pungent odor may also alert an observer to the presence of the triok. The body odor is reminiscent of that of a skunk (*Mephitis mephitis*)—this, combined with bold black and white markings, also makes a strong case for convergence between trioks and skunks.

Semiaquatic and Fossorial Insectivores

Sensory Mucous Glands

The semiaquatic, semifossorial platypus—the sole extant member of the family Ornithorhynchidae, order Monotremata (fig. 2.8A)—inhabits freshwater lakes and rivers along the east coast of Australia and throughout Tasmania. Platypuses generally build simple burrows in stream banks, although those constructed by females for nursing may be 20 to 30 m (65–98 ft) long and branched, with one or more nest chambers at the end. Platypuses are smaller than most people imagine: adult males average about 50 cm (19.6 inches) in length and weigh 1.7 kg (3.7 lb)—about the size of a muskrat. The feet of the platypus are pentadactyl (five-toed), and the manus (forefoot) is webbed; uniquely, this webbing is folded back when the platypus is on land. The physical appearance of the platypus is so unusual that the first specimen, a dried skin taken to London in 1798, was thought to be a fake. It was assumed to be a creation stitched together from the beak of a duck and body parts of a mammal!

The bill of a platypus is soft, pliable, and very sensitive, with nostrils at the tip—quite unlike that of a true duck (fig. 2.8B). The bill is the animal's main sensory organ for navigation and for locating food; it is highly innervated both for tactile reception and to sense electrical fields generated by the muscle contractions of prey (Scheich et al. 1986; Manger and Pettigrew 1995; Proske, Gregory, and Iggo 1998). The small eyes and ears of a platypus are situated in a groove extending from the bill. During a dive, this

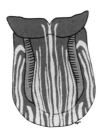

FIGURE 2.8. (A) The platypus (*Ornithorhynchus anatinus*) swimming. (B) Close-up of the head of a platypus. (C) The distribution of large mucous sensory glands (left diagrams) and push-rods (right diagrams) in the platypus bill, viewed from the dorsal (first and third) and ventral (second and fourth) aspects. The large mucous glands are arranged in longitudinal stripes (shaded) alternating with stripes without glands; the push-rods have a more uniform distribution, with a higher concentration toward the edge of the bill. *Sources:* (A, B) David Watts; (C) redrawn by Carie Nixon, Illinois Natural History Survey, from U. Proske and E. Gregory (*Comparative Biochemistry and Physiology A* 136:821–825, 2003), modified from K. H. Andres and M. von During, "The platypus bill," in *Sensory Receptor Mechanisms* (ed. W. Hamann and A. Iggo, World Scientific, 1984).

groove closes and the platypus relies solely on the sensitivity of the bill to locate prey. There are an estimated 40,000 "sensory mucous glands" in the bill that sense electrical fields as a platypus moves its head from side to side while foraging (Proske and Gregory 2003). The skin of the bill is a mosaic of mechanoreceptors (touch receptors) and electroreceptors located on the dorsal and ventral surfaces (fig. 2.8C). The electroreceptors include a series of large mucous glands aligned in longitudinal rows. The mechanoreceptors have a more uniform distribution across the bill and are clustered toward the edge, at the base of a mobile column of flattened, keratinized cells called push-rods (Proske and Gregory 2003). On contact with an object, the column moves upward or downward to stimulate receptors at the base; these touch receptors evoke a reaction in the brain. The system is similar to the way in which movement of a cat's whisker stimulates receptors at its base. As Proske and Gregory (2003: 821) state, much of our information "is speculation, and there is still much to be learned about electroreception in the platypus and its fellow monotreme, the echidna."

Food obtained during a dive is stored in two large internal cheek pouches, opening from the back of the bill. Platypuses feed on a variety of invertebrates, small fish, and amphibians. Smaller prey items are sifted from the substrate by the complex bill apparatus, while larger prey are captured individually. Young platypuses have small, calcified teeth, which are shed before the young emerge from the burrow. Adults lack functional teeth and instead have a series of horny grinding plates and sharp, shearing ridges for most of the length of each jaw. These plates grind food materials before they are swallowed. When eating, a platypus uses its flattened tongue to work against the palate, masticating food while the animal floats on the water surface. Because platypuses are so specialized they are extremely vulnerable to habitat modifications. Many human-induced changes that cause habitat fragmentation in eastern Australia have resulted in local reductions in populations of this fascinating mammal.

Eimer's Organs

Fossorial mammals are equipped with an extensive array of somatosensory receptors, and among the best-studied are Eimer's organs, sensitive tactile organs of moles and desmans of the family Talpidae. Several thousand Eimer's organs are located on the snout (Catania 1995, 1999, 2000; Catania and Kaas 1996). These organs are named after Gustav Heinrich Theodor Eimer, the German zoologist who first described their occurrence in the European mole (*Talpa europaea*) in 1871. The organs differ from one species to another, but the basic structure is the same. Under close scrutiny, Eimer's organs appear like "a mass of bulbous protuberances, reminiscent of a miniature cobbled street" (Gorman and Stone 1990: 47). The organs are found, for example, on the nose of the star-nosed mole (*Condylura cristata*) of eastern North America (fig. 2.9). Each tiny organ is surrounded

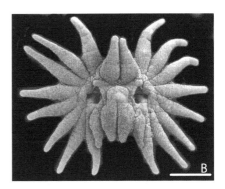
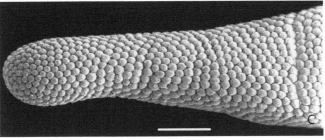

FIGURE 2.9. (A) The star-nosed mole (*Condylura cristata*) has a unique nose with 22 fleshy appendages that encircle the nostrils. (B) Scanning electron micrograph showing each appendage covered with small epidermal papillae called Eimer's organs, which are very sensitive to touch. Scale bar = 2 mm. (C) Close-up of the hundreds of Eimer's organs on each appendage. Scale bar = 250 μm. (D) Drawings of the body parts of a star-nosed mole in their normal anatomical proportions (top) and relative to their proportional representation in the somatosensory cortex (bottom). Note the relatively large somatosensory field for the nose. *Sources:* (A) Kenneth N. Geluso, University of Nebraska, Omaha; (B, C) Kenneth C. Catania; (D) Carie Nixon, Illinois Natural History Survey, adapted from K. C. Catania and J. H. Kaas (*BioScience* 46:584, 1996).

at its base by a blood-filled sinus sitting on a network of sensory nerves; nerve endings pass up from this network into a thick epidermal cap. When the mole touches an object, Eimer's organs rock on their fluid foundations, transmitting the stimulus to the underlying nerve endings via sensory nerves to the central nervous system. The central nervous system stimuli from the organs that have been "disturbed" are received and integrated, thus providing information about the characteristics of the stimulus.

The effectiveness of Eimer's organs in detecting minute surface details was demonstrated by Richard (1982), who trained a tame desman to ascertain which of two boxes contained a food reward, based on a series of miniscule textures engraved on their lids. With the aid of its Eimer's organs, the desman was able to detect engraved lines only one-fifteenth of a millimeter (0.067 mm, or 0.002 inches) in depth and width.

There are two extant species of desmans: the Pyrenean desman (*Galemys pyrenaicus*), weighing about 65 g (2.3 oz) (fig. 2.10), and the larger Russian desman (*Desmana moschata*), weighing 500 g (17.6 oz). Russian desmans inhabit river basins of southwestern Russia and adjacent parts of Ukraine and Kazakhstan. Pyrenean desmans are found in the mountains of southern France, northern Spain, and northern Portugal. The diet of desmans consists of aquatic insects, amphibians, snails, and freshwater shrimp; Russian desmans also consume larger prey such as fish and amphibians. A desman uses its long, sensitive nose to probe in the river mud and between stones and crevices along river banks during its evening foraging bouts. Very well adapted for aquatic life, desmans possess long, fimbriated (fringed with hair) tails, laterally flattened at the tip, that function as a paddle and rudder. Their legs are long and muscular; both the fore and the proportionately enormous hind feet are webbed. When swimming, the hind legs provide the main propulsive force. *Galemys* has been noted to share general body shape and position of hind feet with certain diving birds such as loons (Gaviidae) and grebes (Podicipedidae)—an evolutionary convergence in response to diving activities (Niethammer 1970). The finishing touch in its adaptation for aquatic life is the presence of watertight valvular nostrils and ears.

The ranges of *Galemys* have declined steadily due to pollution of streams and rivers and habitat destruction, compounded by predation by introduced American mink (*Neovison vison*) that escaped from fur farms and became established in the wild. Also, in the late nineteenth century, the Russian desman was intensely trapped for its lustrous fur; at one point,

FIGURE 2.10.   The semi-aquatic Pyrenean desman (*Galemys pyrenaicus*). *Source:* Daniel Heuclin / NHPA/Photoshot.

20,000 skins were processed each year. This species also has declined due to habitat modification and water pollution. Both desman species are listed as endangered.

While the desmans are well adapted for aquatic life, most moles have evolved adaptations for **subterranean** life. However, there are always evolutionary "outliers," and the star-nosed mole (*C. cristata*) of North America is a prime example (fig. 2.9). This mole is unique among mammals in having 22 fleshy, tentacle-like appendages, packed with Eimer's organs, around the tip of its nose. With poorly developed eyesight and only a moderately developed sense of smell, the star-nosed mole uses its impressive snout and long whiskers to probe and search out prey—often underwater. *Condylura* is the only semiaquatic mole, and it prefers water-saturated soils. It can be found in deep, mucky soils of wet bottomlands as well as on steep slopes and in wet areas of high ridges. It often builds tunnels near marshy areas or streams; these tunnels commonly open directly into water. Star-nosed moles inhabiting these muddy substrates are known to hunt along stream bottoms, where they find aquatic invertebrates such as worms and insects—larvae of caddis flies, midges, and stoneflies. During winter they forage below the ice in frozen ponds and streams and are highly dependent on bottom-dwelling prey. *Condylura* is a voracious eater and consumes an amount equivalent to 50% or more of its body mass each day. Its whiskers are located on the sides of the snout, on the sides of the eyes and ears, and even on its forefeet. With so many touch receptors, the star-nosed mole has a better-developed sense of touch than any other mole.

Underwater Sniffing

Several species of shrews have invaded the aquatic world and undertake frequent forays to pursue large invertebrate prey. Convergent evolution has occurred several times in the family Soricidae. Adaptations enhancing a semiaquatic life have developed in four genera—*Sorex, Neomys, Nectogale,* and *Chimarrogale*—inhabiting two different continents.

Water shrews are well adapted for diving and swimming in search of prey (fig. 2.11A). Their snouts are used to probe the underwater substrate; the dense array of whiskers around their nostrils aids in detection of prey, by perceiving shape and texture. These small mammals have a fringe of stiff hairs, or fibrillae, on the lateral edges of the hind feet and toes (fig. 2.11B), as well as on the ventral surface of the tail. The hairs on the feet rise up during the downstroke to increase the surface area of the foot, but fold down and out of the way during the upstroke. As a result, the surface area of the foot is increased at the crucial moment, benefiting propulsion. While the animal is underwater, a fimbriated tail is adaptive in preventing rolling and tends to stabilize body motion while swimming. Perhaps the water shrew best adapted for swimming is the elegant water shrew (*Nectogale elegans*), an inhabitant of montane streams in the Himalayan Mountains and southeastern Tibet. Its fimbriated fore and hind feet are fully webbed, with disklike pads on the base and small scales on the dorsal surface that are helpful in traversing wet stones and perhaps in holding prey. Its nostrils are located behind the nose shield, which may prevent water from entering.

In the New World, three species—the American water shrew (*Sorex palustris*), the marsh shrew (*S. bendirii*), and the Glacier Bay water shrew (*S. alaskanus*)—are dedicated swimmers. The most widespread species is the American water shrew, weighing 10 to 15 g (0.35–0.53 oz), which is found throughout most of North America. As with most shrews, the eyes of *S. palustris* are minute, and the ears are small and hidden under the bicolored coat—jet black above and silver below. *S. palustris* is rarely found far from water. Its nest of dried moss can be found in bankside burrows, under boulders, or along streamside tangles of roots. In eastern North America, *S. palustris* is found in rocky-bottom streams surrounded by

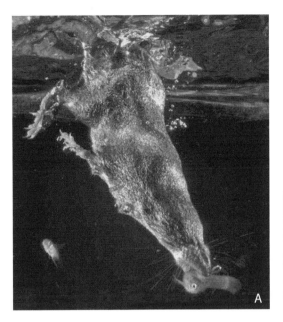

FIGURE 2.11. (A) Foraging in the American water shrew (*Sorex palustris*). (B) Note the fringe of stiff hairs on the margins of the hind foot, shown in a scanning electron micrograph. *Source:* Kenneth C. Catania.

forests of hemlock, spruce, and rhododendron. In the Rocky Mountains I have captured these shrews in boggy meadows of alder surrounded by aspen, spruce, and fir.

Like all other shrews, water shrews are active around the clock, with about 12 one-hour foraging bouts in a 24-hour period. On a given day, these secretive "insectivores" forage excitedly for short periods and then suddenly drop off to sleep. Their diet consists primarily of small aquatic animals such as snails, worms, small fish and their eggs, and insects—including nymphs of caddis flies, stoneflies, and mayflies. Terrestrial invertebrates are also consumed. Interestingly, the diet of water shrews may include large numbers of slugs, snails, and earthworms, as well as the fungus *Endogone*.

Water shrews are active throughout the year, even foraging below the ice during winter. When swimming underwater or crawling on stream bottoms, the water shrew looks like a small silver submarine or a self-propelled bubble. In addition to being an adept underwater swimmer, *S. palustris* can walk or glide on water. One report documented a water shrew running more than 1.5 m (5 ft) across the smooth surface of a pond (Jackson 1961)! This impressive achievement is feasible because the fibrillae can hold small globules of air and act as a sort of hydrofoil.

Even though water shrews are excellent divers, reported to sustain forced dives of up to almost 48 seconds, remaining underwater is difficult (Calder 1969). This difficulty is due to their very dense, water-repellent fur, which does not allow water to penetrate, trapping air bubbles that enhance buoyancy. As a result, the shrew surfaces and floats like a cork whenever it stops paddling.

Perhaps the most spectacular adaptation of the American water shrew is its ability to detect prey underwater by using its sense of smell. Investigations have shown that water shrews can detect odorants while underwater. Recently, Kenneth Catania examined the hunting behavior of water shrews in the laboratory, using a high-speed video system and infrared lighting (Catania 2006; Catania, Hare, and Campbell 2008). With these high-speed video recordings, Catania found that the water shrews continuously emitted and reinhaled air from their nostrils while foraging underwater, indicating that they could be "sniffing" odors while submerged (fig. 2.12). To test this idea, he trained water shrews to follow an underwater scent trail that was randomly laid on either of two paths leading to a food reward. The shrews performed the task with great accuracy. However, when the bubbles they exhaled during underwater sniffing were blocked by a fine steel grid placed over the scent trail, the shrews were unable to follow the scent. As Catania explained, "This ability had been overlooked because the sniffing occurs so quickly it requires slow-motion video to observe, and not many shrews have been filmed underwater with high-speed cameras. I might have overlooked it, too, if I had not already discovered this trick in the star-nosed mole—another semiaquatic mammal that often forages underwater" (Catania, Hare, and Campbell 2008). Both water shrews and star-nosed moles seem to have adapted their olfactory systems for use underwater. While foraging underwater, they exhale air bubbles through their nostrils, often directly onto objects or prey they are investigating.

Subterranean Insectivores

Nearly 75% of all mammalian species are terrestrial or semiaquatic, another 20% (bats) are primarily aerial, and approximately 4% are fossorial,

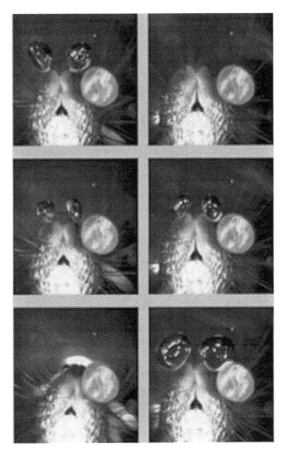

FIGURE 2.12. Laboratory study showing underwater sniffing by the American water shrew (*Sorex palustris*). The bubble is in contact with a wax object. *Source:* K. C. Catania (*Nature* 444:1024–1025, 2006).

spending their life in a subterranean environment, digging under the ground surface to find food or create shelter. This underground environment provides a comparatively simple, closed system where microclimatic factors such as temperature and humidity are predictable and constant and the risk of predation is low. Because of the constancy and microclimatic stability of this environment, much less genetic variation exists in subterranean mammals than in their aboveground counterparts. Convergent evolution is seen in many groups. The subterranean environment has changed little over evolutionary time, and species have optimized survivorship through unique specializations in morphology and behavior. All species have minute ear flaps or nonexistent **pinnae** (external ears); the olfactory, auditory, and tactile senses are well developed. Many have fur coats that are velvety and move easily in both directions, adaptive in their confined burrow systems. Almost all subterranean species are solitary and territorial, often have one litter per year, and produce few young in each litter. Fossorial small mammals include the marsupial moles (family Notoryctidae) of Australia, golden moles (Chrysochloridae) of Africa, and true moles (Talpidae) of the Nearctic and Palaearctic regions, plus about seven different families of rodents from around the world that include the gophers, tuco-tucos, coruros, zokors, mole-rats, voles and mice, and Old World rats and mice (Feldhamer et al. 2007).

Eastern moles (*Scalopus aquaticus*) of North America epitomize the fossorial mammal in form and function (fig. 2.13A). They have short hair, pointed snouts, and torpedo-shaped (fusiform) bodies, and long appendages are replaced with greatly enlarged forelimbs and short stubby tails. The massive forearms are supported and rotated by the teres major, latissimus dorsi, and pectoralis posticus muscles (fig. 2.13C). The forelimbs and pectoral girdle are greatly modified for digging—the large manus with modified sesamoid bone, small clavicle, keeled sternum, elongated scapula (shoulder blade), and the olecranon process of the ulna attached to the massive, rectangular humerus, which articulates with the radius. These morphological adaptations superbly suit eastern moles to their life belowground. Unlike star-nosed moles, the eastern moles tend to avoid wet, loose soils and swamps. Preferring well-drained, sandy soils, the moles make their tunnels in forests, fields, lawns, golf courses, cemeteries, and meadows (fig. 2.13B) Although *S. aquaticus* is often accused of feeding on various garden plants, grasses, and tubers, it in fact eats soil invertebrates, mainly earthworms and insects, both larvae and adults, and only to a lesser degree, vegetable matter (Yates and Schmidly 1978). Eastern moles are active both day and night, year-round. Their tunnel systems are generally of two types: tunnels just beneath the surface and more permanent tunnels deeper underground. The surface tunnels form the characteristic ridges well known to gardeners and golf-course greens keepers. They are used mainly for foraging and are normally dug at a rate of about 3 to 6 m (10–20 ft) per hour. Moles build their tunnels by moving their broad forefeet sideways, using alternating sidestrokes. The soil is then passed under the body to the hind feet and kicked to the rear. Once a pile of soil accumulates behind the mole, it makes a U-turn in its burrow and pushes the dirt, with one of its forefeet, into an unused sec-

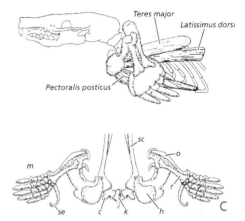

FIGURE 2.13.   (A) The characteristic short hair, pointed snout, fusiform body shape, and large forepaws are evident in this eastern mole (*Scalopus aquaticus*). (B) *S. aquaticus* is a common inhabitant of lawns, fields, and graveyards of eastern and midwestern North America. Surface tunnels and mole hills are common indications of the presence of eastern moles. (C) The massive forearm (top) is supported and rotated by the teres major, latissimus dorsi, and pectoralis posticus muscles. The lower drawing shows the forelimbs and highly modified pectoral girdle of the European mole (*Talpa europaea*). Note the large manus (*m*) with modified sesamoid bone (*se*), the small clavicle (*c*), keeled sternum (*k*), elongated scapula (*sc*), and elongated olecranon process (*o*) of the ulna. The humerus (*h*) is massive and rectangular and articulates with the radius (*r*). *Sources:* (A) Courtesy of The Roger W. Barbour Collection at Elon University; (B) J. F. Merritt; (C) Feldhamer et al. 2007, adapted from M. L. Gorman and R. D. Stone, *The Natural History of Moles* (Cornell Univ. Press, 1990).

tion of the tunnel or to the surface of the ground by way of a vertical tunnel. As a result of these excavations, mole hills are created. The eastern mole is reported to construct up to 31 m (102 ft) of tunnels per day. To repair damaged tunnels, the mole burrows beneath the damaged area and pushes up the floor of the old burrow. The mole's deeper, more permanent tunnels are located about 25 cm (10 inches) belowground. These deeper burrows are commonly dug in spring and are marked only by a circular mound of soil on the ground surface. Within these deep burrows, the eastern mole builds its nest lined with roots, grass, and leaves. The nest chamber measures about 10 to 20 cm (4–8 inches) in diameter and is used as a maternity site as well as a haven from heat and cold. A special location in the complex of tunnels is established for a "latrine."

Seismic Sensitivity

Subterranean environments are not favorable to the use of vision or to detecting airborne sounds as a means of long-distance sensory perception. Seismic vibrations have been shown to travel much better than airborne sound in the underground environment (Narins et al. 1992). The use of seismic signals for communication underground is well documented for rodent species (e.g., members of the family Bathyergidae and mole-rats of the genus *Nannospalax*, family Spalacidae). Many such rodents accomplish this seismic communication through drumming behaviors. For instance, Cape mole-rats (*Georychus capensis*, family Bathyergidae) engage in foot drumming, perhaps to advertise the presence of an occupied burrow to neighboring animals and to convey information on sex and reproductive condition. Other mole-rats, such as the Middle East blind mole-rat (*Spalax ehrenbergi*), generate seismic signals by thumping the roof of the tunnel with their head, in response to vibratory signals from other individuals of the same species (Rado et al. 1987).

Grant's golden mole (*Eremitalpa granti*) of the Namib Desert stands out from the crowd of fossorial, seismic-sensitive mammals: these African moles use seismic sensitivity for the detection of insect prey (fig. 2.14). *E. granti* belongs to the family Chrysochloridae, which encompasses 9 genera and 21 species found throughout central and southern Africa in forests, fields, and plains that have soils suitable for burrowing (Bronner and Jenkins 2005). Many of the same adaptations that enhance under-

FIGURE 2.14.   Grant's golden mole (*Eremitalpa granti*) is a leader among small mammals in monitoring seismic events. *Source:* Galen Rathbun, California Academy of Sciences.

ground movement in marsupial moles (family Notoryctidae) and true moles (family Talpidae) are also seen in chrysochlorids—a prime example of convergent evolution. Grant's golden moles have no pinnae and have poorly developed eyes, with fused eyelids covered with skin. The pelage moves equally well in any direction and is a "metallic" or iridescent red, yellow, green, or bronze, depending on the species. A smooth, leatherlike pad covers the nose, which golden moles use for pushing soil in their subterranean surroundings. Like moles in the family Talpidae, the golden moles' poorly developed eyes cannot form images. Unlike the members of the Talpidae, however, the forelimbs of golden moles are under the body and do not rotate outward. Instead, these moles dig by forward extension of the limbs and scratching at the soil with large claws, especially those on the powerful third digit.

Scattered over the Namib Desert are islands of dune grass and ostrich grass, home to dune termites, the principal insect prey of Grant's golden mole. The termites live among the roots of these clusters of grass. In the presence of light winds that typically blow over the region, the grass clusters emit low-amplitude vibrations that are transmitted through the sand, generating seismic signals nearly 30 dB (decibels) greater than those from adjacent flat areas. Prey-generated vibrations in the grass are also detectable from a shorter range.

When foraging for termites below the surface, *Eremitalpa* propels itself with a sand-swimming motion, the loose sand immediately collapsing behind. *Eremitalpa* is known to orient itself nonrandomly toward the grassy hummocks in which prey items are to be found (Narins et al. 1997; Mason 2003). Moles make remarkably straight paths from grass clump to grass clump; these paths frequently span more than 10 m (32.8 ft) between clumps, and the moles cover an average distance of approximately 1,400 m (0.87 miles) per night without visual aid. Recent work by Lewis and associates (2006) indicates that golden moles may use seismic cues to navigate and locate prey on or beneath the sand. When moving on the surface of the desert, *Eremitalpa* has been observed to occasionally dip its head and shoulders into the sand. Narins and coworkers (1997) suggested that the head-dipping behavior in *Eremitalpa* is a means of tightly coupling the head to the substrate to localize seismic signals. Several authors have suggested that navigation is greatly enhanced in Grant's and Cape golden moles (*E. granti* and *Chrysochloris asiatica*) by the presence of disproportionately large auditory ossicles in the middle ear that serve as adaptations for detecting ground vibrations (Narins et al. 1997; Mason and Narins 2002; Willi, Bronner, and Narins 2006). The combined mass of the **malleus** and **incus** in *Eremitalpa* is approximately 0.1% of the animal's total body mass, compared with 0.00008% in humans.

Aerial Insectivores

Bats are such unusual creatures that some effort is required to think of them as actual animals living in a world of common sense and concrete reality.

—DONALD GRIFFIN (1958)

The most successful group of insectivorous mammals is, without a doubt, the bats (order Chiroptera). Chiropterans are represented by more than 1,100 species worldwide—approximately 22% of all species of mammals (Wilson and Reeder 2005: fig. 1.20). The relationships among chiropteran families are poorly understood, and relationships based on morphological data do not agree with those derived from molecular data. In my discussion of bats, I use the terms *megachiropterans* (or *megabats*) to designate members of the suborder Megachiroptera (family Pteropodidae) and *microchiropterans* (or *microbats*) for members of suborder Microchiroptera (all families other than Pteropodidae). An alternative use of terminology is given by Feldhamer et al. (2007), based on the work of Simmons (2005).

In addition to monopolizing the insect-eating niche, bats have also exploited a wide array of foods ranging from plants, pollen, and nectar, to small vertebrates, and even fish and blood. However, the majority (70%) of microchiropterans are insectivorous (Kunz and Fenton 2003; Patterson, Willig, and Stevens 2003), and all bats living north of 38°N and south of 40°S latitude are insectivorous.

Bats may consume 50% of their body mass in insects each night. A female bat producing milk for her young may consume as much as her own weight in insects in a single night (Kurta 2008). It is possible for a healthy population of bats to consume 4,500,000 insects, or 10 kg (22 lb) of insects, each night. John Whitaker of the Department of Biology at Indiana State University estimates that bats such as the big brown bat (*Eptesicus fuscus*) may eat almost 9 billion insects each year, including many agricultural pests (Whitaker 1995). Freeman (1988) predicted the food habits of free-tailed bats (*Tadarida brasiliensis*) by assessing jaw structure and mechanics; beetle-eaters were characterized by more robust skulls and fewer but larger teeth, whereas moth-eaters had delicate skulls and more numerous, smaller teeth. Jaws vary in depth, presumably in relation to food habits. Long, thin jaws are found in species that feed more heavily on soft-bodied insects, for example. The type of prey ranges from midges and other small swarming flies to large beetles. Contrary to popular belief, though, mosquitoes seldom form more than a minor portion of the diet of bats. Aquatic insects such as midges (family Chironomidae) that form mating swarms are a staple in the diet of little brown bats (*Myotis lucifugus*) and Indiana bats (*M. sodalis*), for example. These swarms provide the bats with natural concentrations of prey and thus are good targets. Most species of mosquitoes, in contrast, do not swarm and are more dispersed. Also, mosquitoes avoid detection by bats in that they tend to hide in the vegetation or very close to it, and thus are essentially unavailable to bats foraging well above the ground.

Many bats of temperate latitudes belong to the family Vespertilionidae (plain-nosed bats). Vespertilionids are small to medium in body size, and in North America they range from about 3 to 30 g (0.1–1.0 oz) in body mass, from the western pipistrelle to the hoary bat. This is the largest family of bats in the world, consisting of 318 species with a worldwide distribution, at elevations up to the tree line.

Bats probably have descended from a rather nondescript shrewlike mammal that pursued insect prey among the branches of trees. It is thought that these diminutive mammals kept to their burrows by day and foraged by night—which would make a great deal of sense, given that they lived side by side with the reigning dinosaurs of North America approximately 60 million years ago. Following an asteroid impact and resultant extinction of approximately 60% of the Cretaceous flora and fauna, natural selection favored mammals capable of flight. This ability permitted access to an almost completely untapped food resource: night-flying insects.

Flight and echolocation work in concert as the premier adaptations enhancing survival for bats. An enduring debate about the course of evolution in modern bats has centered on which of these two adaptations appeared first. Recently, a fossil was discovered in the Green River formation, Wyoming, representing a bat (*Onychonycteris finneyi*) with more primitive features than any other known species. The wing morphology of this 52.5-million-year-old bat is similar to that of extant bat species. However, *Onychonycteris* had a relatively small cochlea, similar to that seen in non-echolocating bats, providing the first direct evidence in support of the hypothesis that flight evolved before echolocation (Simmons et al. 2008).

Wings

Bats are the only mammals capable of powered flight. The ordinal name "Chiroptera," derived from the Greek *cheir*, "hand," and *pteron*, "wing," refers to the modification of the bones of the hand to form a wing, the primary adaptation for flight in bats (fig. 2.15). The wing is formed from skin (**patagium**) stretched between the arm, wrist, and finger bones. The primary modification of the forelimb is its elongation, especially the forearm, metacarpals, and fingers. The radius is greatly enlarged in bats—sometimes being twice as long as the humerus—and the ulna is very much reduced. The thumb is free, while the fifth digit spans the entire width of the wing. The other three digits support the area of the wing between the thumb and fifth digit. The radius cannot rotate (as it does in humans), and the wrist (carpal bones) moves only forward and backward (flexion and extension). Thus, the wing gains strength and rigidity to withstand air pressures associated with flight.

The pectoral region is also highly developed and modified, and this contributes to a bat's ability to fly. The last cervical and first two thoracic vertebrae often are fused. Along with a T-shaped manubrium (the anterior portion of the sternum, or breastbone), the first two ribs form a strong, rigid pectoral "ring" to anchor the wings. In conjunction with the pectoral ring, the articulating scapula and proximal end of the humerus are highly modified for flight. In many species, the **uropatagium** (the membrane between the hind limbs that encloses the tail) also aids in flying. Although not nec-

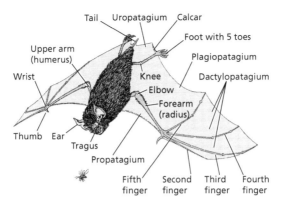

FIGURE 2.15. The major external features of a bat. Note the humerus, the elongated forearm (radius), and the fingers that form the wing. The claw on the thumb (digit 1, or first finger) occurs in all families. Not all species have an external tail and uropatagium. *Source:* Feldhamer et al. 2007, adapted from D. Macdonald, *The Encyclopedia of Mammals* (Facts on File, 2006).

essary for flight (some species have no uropatagium), this membrane may contribute to lift and help stabilize the body during turns and other maneuvers. The size and shape of the uropatagium, and whether it completely encloses the tail, varies considerably among and within families. The **calcar**, a cartilaginous process that extends from the ankle, helps support the uropatagium. Aerodynamic stability is enhanced by a bat's body mass being concentrated close to its center of gravity (Neuweiler 2000).

The hind limbs of bats are small relative to the wings and are unique among mammals: the limbs project sideways and backward, and the knee bends back rather than forward as in other mammals; the legs are adapted for pulling rather than pushing. The lower leg (tibia) is formed by a single bone. This aids in various flight maneuvers, as well as in the characteristic upside-down roosting posture of bats, in which they hang by the claws of the toes.

Compared with birds, flight in bats is slow but highly maneuverable (for detailed analyses of flight in vertebrates, see Norberg 1990). A bat (like a bird or an airplane) is able to fly because it generates enough lift to overcome gravity and sufficient forward propulsive thrust to overcome drag (see Feldhamer et al. 2007: fig. 13.3). Lift and thrust are achieved through the structure of the bat's wings, in conjunction with the pectoral bones and muscles, the uropatagium, and the hind limbs. The dorsal surface of the wing in bats, as in airplanes, is convex, and the ventral surface is concave, which causes air to move more rapidly over the wing than underneath it. This reduces the relative air pressure above the wing and results in lift. In addition, to facilitate airflow, the surface of the wing is kept smooth and taut by layered elastic tissue in the wing. Generally, the greater the camber (the extent of front-to-back curvature of the wing), the greater the lift that can be produced. Likewise, the greater the angle of attack, the greater the lift that can be generated. If the camber and angle of attack are too great, however, the smooth flow of air over the surface of the wing is disrupted and becomes turbulent; lift is greatly reduced or lost completely. Bats are capable of quickly adjusting both camber and angle of attack throughout the wing-beat cycle by use of several muscles in the wing, in conjunction with the movement of the wrist, thumb, fifth finger, and hind limbs.

The shape of a bat's wing is another factor that influences aerodynamic properties. When viewed from above, wing shape varies from short and broad to long and thin in different families and species (fig. 2.16). The proportion of wing length to width is the "aspect ratio." A high aspect ratio—long, narrow wings—adapts a species for sustained, relatively fast flight, as is seen in many free-tailed bats. These species would be expected to forage high above the ground, predominately in open habitats, free of obstructing vegetation. A low aspect ratio—short, wide wings—is seen in bats with slower, more maneuverable flight, as in the false vampire bats and the slit-faced bats. A low aspect ratio is seen in bats that forage more often in habitats with dense, obstructing, understory vegetation. Slow and maneuverable flight can also be enhanced by low "wing loading capacity"—the ratio obtained by dividing the body mass of the bat by the total surface area of the wing. The lower the wing loading, the greater is the potential lift and capacity for slow flight. In general, a direct relationship exists between the wing morphology and flight patterns of bats and their foraging and life history characteristics (Altringham 1996).

Echolocation

But how, if God love me, can we explain or even conceive in this hypothesis of hearing.

— LAZARO SPALLANZANI (1729–1799)

The vocal repertoire of bats encompasses a wide variety of sounds, including those that are low in frequency (<20 kHz [kiloherz]), within the range

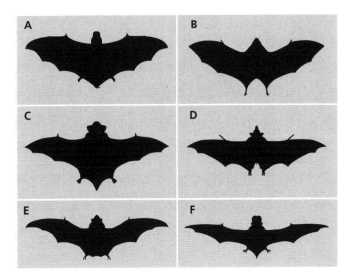

FIGURE 2.16. Increasing aspect ratios of the wings in several families of African bats. The aspect ratio increases as wings become longer and thinner. Broader wings generally translate to slower, more maneuverable flight patterns. (A) Egyptian slit-faced bat (*Nycteris thebaica*) (family Nycteridae); (B) heart-nosed bat (*Cardioderma cor*) (Megadermatidae); (C) Moloney's flat-headed bat (*Mimetillus mononeyi*) (Vespertilionidae); (D) straw-colored fruit bat (*Eidolon helvum*) (Pteropodidae); (E) Pel's pouched bat (*Saccolaimus peli*) (Emballonuridae); (F) Midas's free-tailed bat (*Mops midas*) (Molossidae). *Source:* Feldhamer et al. 2007, adapted from J. Kingdon, *East African Mammals,* vol. 2A (Univ. of Chicago Press, 1974).

of human hearing. Such vocalizations facilitate communication about territorial spacing, recognition, warnings, and interactions between females and their young. However, for hunting prey, the majority of bats produce ultrasonic sound pulses (<20 kHz) and discern crucial information from the echoes that return (fig. 2.17A). The sound pulses are produced in the larynx and emitted from the nose or mouth, traveling at approximately 340 m/second. For many species, a nose leaf acts as an acoustic lens, focusing the sound into a narrow beam in front of the bat. When reflected, the sound pulse that returns is altered according to the physical features of the object it met. Bats can discriminate echo delays as short as seventy-millionths of a second, and many species can determine the presence of objects as thin as 0.06 mm (0.002 inches) in diameter, the width of a human hair.

The discovery of this fascinating adaptation is a journey that has spanned centuries. Serious scientific inquiry into the navigational ability of

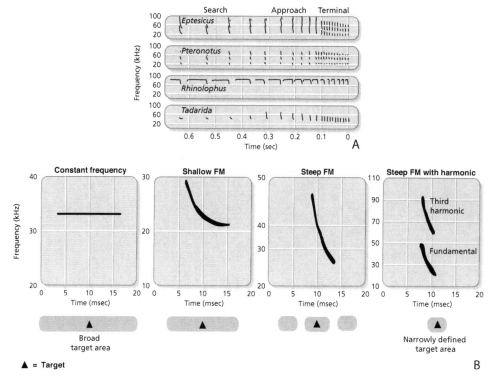

FIGURE 2.17. The dynamics of echolocation. When sounds emitted from the bat's mouth come into contact with objects in the environment, the sound waves bounce back to its ears in the form of echoes. The bat's brain then interprets these sound waves as three-dimensional pictures of the objects present in its flight path. (A) Sonogram showing the increased pulse repetition at the terminal phase of echolocation for several bat genera. (B) Typical pattern of frequency changes in echolocation pulses that continuously narrows the target area perceived by a bat. Bats may use a narrow-band, long-duration, constant-frequency pulse until a potential prey item (represented by the solid triangle) is located. The target area (thick shaded line) perceived by the bat becomes more and more narrowly defined as the bat shifts to broad-band, shorter-duration, echolocation pulses, including a shallow frequency-modulated (FM) sweep, a steep FM sweep, and finally a steep FM sweep with harmonics (integral multiples of fundamental frequencies), which pinpoint the target and provide fine details. *Source:* Feldhamer et al. 2007.

bats began in the 1790s when Italian scientist Lazaro Spallanzani conducted a series of experiments to determine how bats were able to orient in darkness (Galambos 1942; Dijkgraaf 1952). Spallanzani tied hoods of opaque material around the heads of his chiropteran subjects and found that the bats were rendered helpless in flight. However, to Spallanzani's surprise, even material transparent enough for the bats to see through caused disorientation. Furthermore, after damaging the eyesight of some bats, he noted that a blinded bat "shows himself just as clever and expert in his movements in the air as a bat possessing its eyes." This discovery led Spallanzani to believe that bats relied on an undiscovered organ or sense for navigation. In 1794, Charles Jurine, a Swiss colleague of Spallanzani, began to conduct his own experiments, which led to the crucial discovery that if the ears of a bat were obstructed with wax or other material, the bat collided helplessly with obstacles (Jurine 1798). Support from the Jurine experiments inspired Spallanzani to pursue additional studies. By inserting brass tubes into the ears of bats, he found that the bats could orient only when the tubes were open; when the tubes were blocked, the bats could not orient and blundered about helplessly. Unfortunately, the idea Spallanzani and Jurine arrived at—that bats could use their ears to detect stationary obstacles—seemed absurd at the time and was rejected by nearly all zoologists for more than a century.

The details regarding how bats orient and detect prey remained a mystery until the twentieth century, when the puzzle was revisited and resolved. In 1938, Donald Griffin, an undergraduate student at Harvard, approached physicist G. W. Pierce and suggested that they listen to some bats with a then state-of-the-art ultrasonic detector. As a result, the chorus of high-frequency sounds emitted by bats was heard by humans for the first time. Griffin coined the term **echolocation** and hypothesized that bats emit the high-frequency sound pulses from the mouth and use the ears to gather the returning echoes in order to navigate in darkness (Griffin 1958).

The remarkable aspects of echolocation have been studied by numerous investigators following Griffin's pioneering work in the late 1930s. The sound pulses, produced by contraction of the cricothyroid muscles of the larynx, generally span the range of 15 to 100 kHz. Bats perceive a great amount of detailed information from returning echoes, including size, shape, texture, and relative motion, in addition to distance to a target. When a bat is capturing insects in flight, the frequency and characteristics of echolocation pulses vary, depending on which phase of call speed the bat is using: the search, approach, or terminal phase (fig. 2.17A). The search phase is typically composed of constant-frequency (CF) sound pulses. These sounds are of constant pitch and are particularly useful for discriminating between moving and stationary objects. Moving objects can be determined by their

Doppler shift, which is detected when the echo received has changed frequency compared with the sound pulse the bat originally emitted. This change in frequency is caused by the movement of one or both objects: the source (bat) and/or the target (insect). Bats may compensate for these shifts by adjusting the frequency of their subsequent calls, a behavior known as Doppler-shift compensation (Schnitzler 1972). This ensures that the received echoes will remain within a narrow range of frequency in order to stimulate a region of the cochlea that is innervated by a disproportionately large population of neurons with exceptionally sharp tuning properties, the "auditory fovea" (Schuller and Pollak 1979). This behavior represents one of the most precise forms of sensory-motor integration known.

Frequency-modulated (FM) sound pulses incorporate a broader band of frequencies and are better than CF pulses for determining the finer details of a target. Thus, bats may use a series of FM pulses when advancing closer to a target during the approach and terminal phases. During the terminal phase of echolocation, bats also employ harmonics—integral multiples of fundamental frequencies (fig. 2.17B).

In addition to the frequency changes that occur throughout a bout of foraging, the rapidity with which the sound pulses are produced is also modified. During the search phase, the pulse rate is slowest, about 25 pulses/second; for the duration of the approach phase, the rate doubles to about 50 pulses/second. During the terminal phase, just before attack, bats shift to their most sensitive pulse rate—250 pulses/second. This terminal phase is commonly called a "feeding buzz" and contains highly detailed information about the insect. The search-approach-termination sequence of echolocation is repeated very rapidly.

This general description of echolocation applies to most insectivorous bats; however, there is substantial variation and diversity according to foraging situation and species (Altringham and Fenton 2003). For example, some insect-eating bats that fly and feed among vegetation use low-intensity (60–80 dB) calls. These are called "whispering bats" and include slit-faced, New World leaf-nosed, and false vampire bats, and many mouse-eared and long-eared bats. Their much lower-intensity signals have a less effective range than is found in species using signals of higher intensity, and these "whispering" species do not use echolocation for detecting prey but, instead, listen for sounds made by prey's fluttering wings or footfalls.

The tremendous diversity in size and shape of external pinnae among insectivorous bats (see Altringham 1996: 83) reveals the crucial nature of hearing in this group of small mammals. The inside surface of the pinnae is frequently decorated with folds or ridges and may be covered with bands of hair in some species. Most microchiropterans have a small flap at the base of the ear, called a **tragus**, which is important in echolocation and is variable in size and shape. The size of ears varies greatly in the microbats, from the tiny cup-shaped ears of the sac-winged and mouse-tailed bats, to the funnel-shaped ears of the funnel-eared and disk-winged bats, to the remarkably long ears of false vampire and slit-faced bats.

Regardless of the size or shape of the ear, once a sound wave enters the pinna, it strikes the tympanic membrane and is converted to mechanical vibrations. These vibrations are amplified and transmitted via the three mammalian ear ossicles (stapes, incus, and malleus) to the oval window of the inner ear. Fluid in the inner ear concentrates and transmits vibrations to the basilar member, where vibrations are converted to neural impulses that pass to the brain. The intensity of sound pulses emitted by bats may be 120 dB—analogous to a blaring smoke detector. To avoid self-deafening by their own voice, bats accomplish a disarticulation in the middle ear. The stapedius muscle attaches to the **stapes;** just before a bat emits an echolocation pulse, this muscle contracts and pulls the stapes away from the oval window. As a result, sound is prevented from reaching the cochlea, a very delicate structure that can be damaged by loud sounds. As soon as the pulse has been emitted, the bat relaxes its stapedius and its hearing is fully restored.

The work of J. E. Hill and Smith (1992) with Brazilian free-tailed bats (*Tadarida brasiliensis*) revealed that contraction of the stapedius muscle occurs 10 milliseconds before generation of the vibration pulse, pulling the stapes away from the oval window and preventing emitted sounds from reaching the inner ear. In only a few milliseconds, the stapedius relaxes and reengages the stapes with the oval window to permit normal detection of the echo. In the final stages of taking an insect, a bat's sonar pulse repetition rate may exceed 200 pulses/second, meaning that the stapedius muscle may operate at the same frequency; this is one of the highest contraction-relaxation rates recorded in vertebrate muscle.

Another sound-dampening specialization involves the bony otic capsule that houses the middle ear. Instead of being fused to the skull, as is common in mammals, the otic capsule in bats is loosely suspended in its cavity and is surrounded by blood sinuses and fatty deposits. This lack of adhesion is crucial in minimizing conduction of sounds from the larynx and respiratory passages through the bones of the skull.

Although bats have developed the most sophisticated use of echoloca-tion in the class Mammalia, some other mammals emit ultrasonic sounds: seals, shrews, tenrecs, rodents, and a few marsupials. The toothed whales (suborder Odontoceti) employ an echolocation system that is similar to the sonar systems used by submarines. Some animals are known to echolocate with lower-frequency sounds, such as South American oilbirds (*Steatornis caripensis*) and the cave swiftlets (*Collocalia*).

Most megabats rely on vision and olfaction to locate food, avoid obsta-cles, and maneuver in their environment (although the rousette fruit bats, *Rousettus*, have a limited form of echolocation using low-frequency tongue clicks). Recent research by Holland and colleagues has demonstrated that in addition to their ability to orient, navigate, and hunt by echolocation, bats also utilize the earth's magnetic field for navigation (Holland, Waters, and Rayner 2004). The homing ability of big brown bats (*E. fuscus*) was al-tered by artificially shifting the earth's magnetic compass.

Sadly, many species of bats are experiencing declines in population numbers. For instance, Brazilian free-tailed bats (*T. brasiliensis*) are high-altitude, fast-flying members of family Molossidae; they are known to form the largest aggregations of mammals in the world (McCracken 2003). Re-cent research by Betke and colleagues (2008), using thermal infrared imag-ing techniques, reveals that roughly 4 million Brazilian free-tailed bats re-side in six major cave colonies in the southwestern United States. This estimate, based on cave emergence flights, is impressive; however, it reflects a marked reduction from the estimates of colony size of 54 million obtained in 1957 using visual observations (Constantine 1967). The discrepancy be-tween the two censuses may be due to a combination of overestimation in the original study and population decline caused by food-chain poisoning, vandalism, and other causes.

Another recent decline in bat populations has been unfolding in north-eastern North America. A malady known as white-nose syndrome has killed tens, perhaps hundreds, of thousands of hibernating bats of at least five spe-cies in the northeastern United States since 2006. Alarmingly, mortality rates exceeding 90% have been reported in some **hibernation** caves. Dead or dying bats are typically dehydrated and emaciated, left with little or none of the fat stored to survive the long months of hibernation. In late 2008, sci-entists reported that the causal agent for the disease may be an unusual form of *Geomyces* fungi, which are usually found in cold places such as Antarctica (Blehert et al. 2008). Because such fungi do not usually kill otherwise healthy animals, it is thought that the infection may cause a bat to arouse too often during hibernation, thus causing a depletion of its fat reserves in midwinter. If the mystery of white-nose syndrome remains unsolved, this could result in an ecological calamity.

CHAPTER THREE

Herbivory

....

The most abundant foods on earth are plants and insects, so it is not surprising that the most abundant small mammals are rodents and bats. Rodents constitute the largest mammalian order, with 2,272 species—roughly
42% of all extant mammalian species. The radiation of rodents over the
course of evolution is evident in the diversity of their unique **adaptations**.
The order Rodentia has a cosmopolitan distribution, with species occupying virtually every type of habitat, ranging from the high arctic tundra to
the hottest and driest deserts. Rodents have a diverse array of locomotor
adaptations, permitting them to spend their lives underground or in the
canopy of forests, where they glide from tree to tree. Some live entirely on
seeds, with little or no water, while others have webbed feet and a semiaquatic lifestyle. Others are dependent on humans, such as the **commensal**
house mice and rats. The majority of rodents are small (20–100 g, or 0.07–
3.5 oz), although the largest, the capybara (*Hydrochaerus hydrochaeris*),
may reach 50 kg (110 lb). Despite the large number of species and their
widespread distribution and diversity, rodents are surprisingly uniform in
several general morphological characteristics.

Herbivorous ("plant-eating") mammals consume green plants and thus
form the base of the consumer food web. As early-twentieth-century ecologist Charles Elton (1927: 57) pointed out, animals that feed on plants can
be viewed as the "key industry" in the ecosystem—a large number of animals depend on these small mammalian herbivores for sustenance. Plant
food is far more abundant than animal food, but its energy content is lower.
Gaining access to the protein in leaves and stems is difficult, given the tough
fibrous cell walls of plants.

We can divide herbivores into two main groups: (1) browsers and grazers, such as the hoofed mammals—the orders Perissodactyla and Artio-

dactyla; and (2) the gnawers—the orders Rodentia and Lagomorpha. Other herbivores are the kangaroos, wallabies, wombats, langurs, sloths, elephants, and hyraxes, and aquatic grazers such as manatees and dugongs. Herbivores feed on a great variety of foods, including grasses, leaves, fruits, seeds, nectar, pollen, and even the sap, resins, or gums of plants.

General Characteristics

The jaw muscles of herbivores differ from those of **carnivorous** animals (fig. 3.1). The jaw movement in mastication is from side to side (not up and down, as in carnivores), and the upper cheekteeth slide across the complementary surfaces of the lower teeth in a sweeping motion. For herbivores, the major muscles involved in mastication are the **masseter** and **pterygoideus**, and the **temporalis** muscle is smaller than in carnivores. The distance between the lower **incisors** and cheekteeth offers the advantage of allowing for a narrow snout that can penetrate into small spaces to crop food, as seen in smaller deer, antelopes, and rodents.

Ever-Growing Incisors

Many herbivores share unifying characteristics in the design of the skull and teeth. In general, herbivorous mammals are typified by skulls in which the **canines** are reduced or absent and broad **molars** are adapted for crushing, shredding, and grinding fibrous plant tissue. A defining characteristic of dentition in lagomorphs is an additional pair of "secondary" upper incisors located immediately behind the first pair. The characteristic that defines all rodents is a single pair of large upper and lower incisors, which are open-rooted, ever-growing, and used for gnawing. Gnawing quickly wears down the tips of the incisors and a chisel-like edge forms, because the anterior side of each incisor is covered with enamel and wears more slowly than the posterior side, which lacks enamel. The incisors are used to gnaw through hard plant coverings to reach the tender material inside, as well as for nibbling grasses and shrubs. In herbivores, three main masticatory muscles (the temporalis, masseter, and pterygoideus) regulate how these animals shred and grind tough, fibrous food. The arrangement and functions of these muscles are responsible for the forward-and-backward jaw movements of rodents, in contrast to the lateral chewing movements of lagomorphs. The lips of ro-

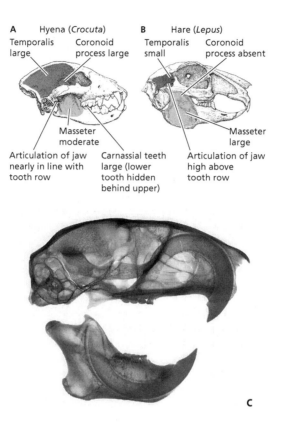

A Hyena (*Crocuta*)
Temporalis large   Coronoid process large

B Hare (*Lepus*)
Temporalis small   Coronoid process absent

Masseter moderate

Masseter large

Articulation of jaw nearly in line with tooth row

Carnassial teeth large (lower tooth hidden behind upper)

Articulation of jaw high above tooth row

C

FIGURE 3.1.   Skulls and teeth of herbivorous and carnivorous mammals. (A) A hyena (*Crocuta*) has specializations adaptive for carnivory, such as a pronounced temporalis muscle, large, conical canines, and carnassial teeth used for shearing flesh. (B) A hare (*Lepus*), a herbivore, has an enlarged masseter muscle. (C) X-ray of the skull of a southern Asian tufted ground squirrel (*Rheithrosciurus macrotis*), showing the pronounced ever-growing incisors found in all rodents. *Sources:* (A, B) Feldhamer et al. 2007; (C) R. W. Thorington Jr. and K. Ferrell, *Squirrels: The Animal Answer Guide* (Johns Hopkins Univ. Press, 2006).

dents can be folded in behind the incisors to prevent chips of bark or soil entering the mouth during gnawing. This adaptation is especially obvious in the semiaquatic beavers and **fossorial** kangaroo rats, gophers, and mole-rats. Rodents also have a **diastema**, a gap between the incisors and cheekteeth that allows for maximum use of the incisors in manipulating food. As in other herbivores, a diastema posterior to the incisors results from the absence of canine and **premolar** teeth. Food can also be held in the diastema before it is passed back to the cheekteeth for processing—this is most obvious in grass-eating herbivores. The number of molar teeth is reduced. Rodents have an impressive array of molar dentition, with the **occlusal** surface arranged in convoluted layers of enamel adapted for various diets. Molars may or may not be ever-growing. In rodents, the number of teeth is greatly reduced from the primitive eutherian number, typically 16 teeth in all. Rodents never have more than two pairs of premolars, and no species has more than 22 teeth—except the silvery mole-rat (*Heliophobius argenteocinereus*) of central and east Africa, which has 28. Rodents generally are herbivorous or **omnivorous**, depending on the season and availability of food items. The jaw musculature and skull structure associated with the dentition serve as important criteria for grouping rodents.

Coprophagy

Digestion of cellulose occurs in the **cecum** (a blind sac, or appendix, in the digestive tract, located between the small and large intestines) in 11 families of rodents, lagomorphs, hyraxes, and the **marsupial** common ringtail, to mention just a few. Because there is no regurgitation and the rate of passage of forage is rapid, these mammals process a minimal amount of the fiber when they first ingest plants. As a result, **coprophagy** (refection), the feeding on feces, has evolved in lagomorphs, rodents, shrews, and some marsupials (fig. 3.2) (Altuna, Bacigalupe, and Corte 1998; Hirakawa 2001). The cecum houses bacteria that aid in digestion of cellulose. Most products of digestion pass through the gut wall into the bloodstream, but certain nutrients such as essential B vitamins produced by microbial fermentation do not. Such vitamins and minerals would be lost if lagomorphs did not eat some of their feces and so pass this material through the gut twice. To optimize the uptake of essential vitamins and minerals and enhance assimilation of energy, lagomorphs produce two kinds of feces. The first type is moist, mucus-coated, cecal pellets, excreted and promptly eaten directly from the anus. These are stored in the stomach and mixed with food derived from the alimentary mass. The second type is hard, round feces that are passed normally. The frequency of coprophagy in rabbits is usually twice daily. Prevention of coprophagy in laboratory rats resulted in a 15% to 25% reduction in growth.

The mountain beaver (*Aplodontia rufa*) of Pacific Northwest America is a classic coprophagist (fig. 3.3). The name "mountain beaver" is rather inappropriate, as this species is not restricted to mountains and is not re-

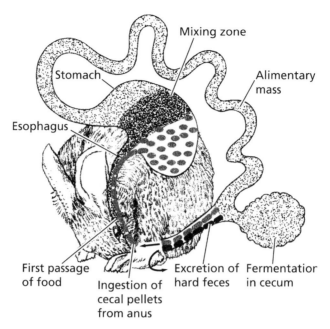

Mixing zone

Stomach

Alimentary mass

Esophagus

First passage of food

Ingestion of cecal pellets from anus

Excretion of hard feces

Fermentation in cecum

FIGURE 3.2. Coprophagy occurs in shrews, rodents, and lagomorphs. The digestive tract is highly modified for coping with large quantities of vegetation. The alimentary canal has a large cecum, which contains bacterial flora to aid in the digestion of cellulose. *Source:* Feldhamer et al. 2007, adapted from D. Macdonald, *The Encyclopedia of Mammals* (Facts on File, 2006).

lated to beavers. Weighing approximately 0.9 kg (2 lb), *A. rufa* is considered the most primitive living rodent. It is strictly herbivorous, consuming plants that other mammals find toxic or unpalatable, such as ferns, stems of Douglas fir, rhododendron, and even devil's club and stinging nettle. *Aplodontia* is a "hindgut fermenter" (discussed in more detail later in the chapter). To optimize extraction of nutrients from plants, it **reingests** soft fecal pellets. When defecating, it sits on its rump with its tail protruding upward and anus partially everted. With an upward jerk of the head, it grabs fecal pellets with its incisors as the pellets emerge from the anus. Next, it "tosses" the hard pellets onto the fecal pile; the soft pellets are swallowed. The hard pellets are reingested at a later time. Normally, the process takes two to five minutes. (For a detailed description of coprophagy in sewellels, see Ingles 1961.) Fecal chambers (or caches) are located both above and

FIGURE 3.3. A mountain beaver (*Aplodontia rufa*) defecating and reingesting fecal pellets. *Source:* L. G. Ingles (*Journal of Mammalogy* 42:411, plate I, 1961).

below ground. Mountain beavers construct extensive burrow systems, and as many as six pellet chambers have been found in a single burrow system, with pellets in different states of decomposition.

Coprophagy has been documented in nine species of shrews, including the northern short-tailed shrew (*Blarina brevicauda*). The adaptive significance of coprophagy for shrews is unknown, but it may represent a technique for reducing daily food intake and, as noted above, extracting certain essential nutrients and vitamins from the available food.

Granivores

Among mammals, granivory (**granivorous** means "seed-eating) seems to be nearly exclusive to small rodents; however, as always, there are exceptions—one of which is the Sechuran fox (*Lycalopex sechurae*). This small fox of northwestern Peru is known to consume primarily seeds; however, it is somewhat opportunistic and is known to eat grasshoppers and rodents at certain times of the year. Small mammals constitute the main granivorous taxa in deserts of North America and North Africa and in Asia. The best known of these are the heteromyids (kangaroo rats and mice and pocket

FIGURE 3.4.   Three species of desert-dwelling granivorous rodents. (A) Arizona pocket mouse (*Perognathus amplus*); (B) Ord's kangaroo rat (*Dipodomys ordii*) (family Heteromyidae); and (C) small five-toed jerboa (*Allactaga elator*) (Dipodidae). *Sources:* (A, B) T. L. Best; (C) Andrey Tchabovsky.

mice) of North America (fig. 3.4A, B). Members of this family (Heteromyidae) are reported to remove more than 75% of the seed bank (J. F. Nelson and Chew 1977; Brown, Reichman, and Davidson 1979). Jerboas (family Dipodidae) inhabit the region from the Sahara Desert eastward across southwestern and central Asia to the Gobi Desert (fig. 3.4C). They are the principal granivores of Old World deserts and steppe environments. In response to **xeric** environments, many of the same adaptations have evolved in jerboas as in the heteromyids, such as long feet and tufted tail, reduced size of forelegs, bipedal gait, large eyes, sandy-colored **pelage**, and the ability to conserve water. Gerbils (family Muridae) live in the deserts of North Africa and southwest Asia and are also principal consumers of seeds. In contrast, ants seem to be the main granivores in deserts of South America, Australia, and South Africa (Mares and Rosenzweig 1978; Morton 1985;

Kerley and Whitford 1994). Australia's rodent granivores have a much more varied diet than do the North American heteromyids (Murray and Dickman 1994). For example, for two Australian murids (spinifex hopping mouse, *Notomys alexis,* and sandy inland pseudomys, *Pseudomys hermannsburgensis*), seeds are a major component in their diet; however, in autumn, these murids focus on invertebrates, which make up some 50% to 60% of food eaten at this time of year.

Food Hoarding

During autumn in seasonal environments, certain mammals are commonly observed harvesting fallen nuts and establishing food stores in preparation for winter—squirrels are a familiar example. This food hoarding, or **caching,** has been reported for members of 6 orders and 30 families of mammals (Vander Wall 1990). Caching confers the advantage of providing a reserve food supply during lean periods. Further, the larder is isolated from competitors and offers protection from predators during feeding. Caching techniques, location, composition, and season, and how food is used, vary at both the individual and species levels. Caching is performed either by "larder hoarding" (concentration of all food at one site) or by "scatter hoarding" (one food item stored at each cache site). Vander Wall (1990) documented the groups of mammals that store food and summarized their food-storing behavior. Most mammalian hoarders are in the order Rodentia; however, caching also occurs in other groups of mammals such as the orders Carnivora and Soricomorpha Only one marsupial, the pygmy possum (*Burramys parvus*), is reported to store food. (My discussion here of this fascinating topic of food hoarding extends beyond the herbivorous small mammals.)

Of the 376 species of shrews, only 7 cache food. This behavior has been documented primarily in the laboratory with food provided *ad libitum.* Among shrews, the premier scatter-hoarder is the northern short-tailed shrew (*B. brevicauda*). Foods cached by *Blarina* include beechnuts, earthworms, insects, snails, plant material, and even small mice and voles. As mentioned in Chapter 2, the venomous saliva produced by *Blarina* plays an important role in preserving caches of prey in a "fresh" state for several days. Such underground caches may be numerous and important in optimizing overwinter survival of the northern short-tailed shrew (Merritt 1986).

European moles (*Talpa europaea*) collect and cache earthworms and insect larvae. After mutilating the head segments of earthworms, the moles cache the worms in chambers and in the walls of galleries near the nest as food for the winter. An interesting aside: when soil temperature increases in the spring, some of the remaining worms, trapped in the galleries, may regenerate a head and burrow to escape (Gorman and Stone 1990).

The most common hoarders are rodents. They store food in many different locations, usually for periods longer than 10 days. Larder- and scatter-hoarding techniques are employed to cache foods ranging from seeds and nuts to woody vegetation, roots, and invertebrates (Steele and Koprowski 2001). As winter approaches, eastern chipmunks (*Tamias striatus*) carry large amounts of food in their cheek pouches and cache the food in their burrows for winter use. Preferred items in their winter diet include hickory nuts (*Carya*), beechnuts (*Betula*), maple seeds (*Acer*), acorns (*Quercus*), and a long list of seeds of woody and herbaceous plants. Gray squirrels (*Sciurus carolinensis*) and fox squirrels (*S. niger*) are principal consumers of the acorns of red and white oaks in eastern North America (fig. 3.5A, B). Steele and associates (2006) have demonstrated the importance of embryo excision (removal of the embryo from seed) as a means of long-term cache management by gray squirrels. Red oaks (*Quercus rubra*) exhibit delayed germination of acorns, and these acorns can be stored up to six months before they begin to germinate. Acorns of white oaks (*Quercus alba*) show no dormancy and germinate in autumn soon after the seeds fall; however, if the embryo is excised, these acorns will remain intact for up to six months. Gray squirrels were found to cache significantly more acorns of red oak species than of white oak species after excision of the embryos. Squirrels excised the embryos of red oaks only when the acorns began to germinate following winter. Naive captive gray squirrels—that is, those without previous experience with acorns—also cached red oak acorns in preference to white oak acorns and attempted embryo excision on white oaks, suggesting a strong innate tendency for this behavior. Such excision attempts, however, were often unsuccessful, indicating that the behavior is probably perfected through learning (Steele et al. 2006). Flying squirrels (*Glaucomys volans*) are more selective than eastern chipmunks; hickory nuts may comprise up to 90% of the nuts they store for winter. Unlike chipmunks (*Tamias*) and tree squirrels (*Sciurus*), red squirrels (*Tamiasciurus hudsonicus*) and Douglas squirrels (*T. douglasii*) (fig 3.5C, D) establish large surface **middens** to cache conifer cones and mushrooms.

Rodents of the family Heteromyidae primarily cache seeds, but some

FIGURE 3.5. (A) The eastern gray squirrel (*Sciurus carolinensis*) and (B) eastern fox squirrel (*Sciurus niger*) are common tree squirrels of North America that practice scatter hoarding. (C) The North American red squirrel (*Tamiasciurus hudsonicus*) is an avid collector of conifer cones. (D) It stores the cones by larder hoarding in centralized locations referred to as middens. *Sources:* (A–C) Hal S. Korber; (D) National Park Service, in R. W. Thorington Jr. and K. Ferrell, *Squirrels: The Animal Answer Guide* (Johns Hopkins Univ. Press, 2006).

species also store fruits, dried vegetation, and even fungi (Reichman and Rebar 1985). Equipped with large, external, fur-lined cheek pouches and a keen sense of smell, heteromyid rodents are the most specialized seed-eaters. The diversity and availability of seeds in desert ecosystems is key to the evolutionary success of these seed-eating desert mammals. In terms of the biomass of seeds harvested, kangaroo rats and allies are rivaled only by ants as important granivores inhabiting North American deserts. Rodents are reported to use more than 75% of all seeds produced at certain Mojave and Chihuahuan Desert sites. In the Mojave Desert of California, Merriam's kangaroo rat (*Dipodomys merriami*) was found to consume more than 95% of the seeds produced by the annual *Erodium cicutarium*. Maximum numbers of seeds produced in desert habitats of North America range from 80 to 1,480 kg/ha (1 ha [hectare] is 10,000 m^2, or 2.47 acres). Minimum densities of seeds remaining in the soil years after the last seed crop are rarely below 1,000 seeds/m^2. As a result of the abundant seed resources and competition with ants, birds, and other rodents, heteromyids have evolved fascinating morphological and behavioral adaptations to optimize their foraging success.

Seed caching is not limited to squirrels and desert rodents. Some small mammals, as noted above, store food in scattered surface caches and underground chambers, whereas others concentrate their cache in a single or just a few larder sites. These foods are consumed during the winter when food supplies are scarce. Eastern woodrats (*Neotoma floridana*) gather and store large quantities of fruits, seeds, leaves, and twigs in their large surface dens constructed of sticks (Wiley 1980). Perishability and nutrient content of food influence the caching decisions of woodrats (family Cricetidae) (Post, Reichman, and Wooster 1993; Post et al. 2006) and kangaroo rats (Heteromyidae) (Price, Waser, and McDonald 2000). Greater blind mole-rats (*Spalax microphthalmus*) of the steppes of Eurasia are noteworthy for establishing many storerooms up to 3.5 m (11.5 ft) in length, which they pack with rhizomes, roots, and bulbs. These large caches are essential for meeting the energy demands of mole-rats during the long winter on the steppes, when the ground surface is frozen and foraging is restricted. Among lagomorphs, only the pikas (*Ochotona*) establish food caches, called hay piles (fig. 3.6). Hay-gathering behavior is common to most pika species. *Ochotona princeps* of North America collects green vegetation during late summer and autumn and establishes hay piles under overhanging rocks in talus (rubble or scree) deposits. Hay piles function as a source of food during winter, as well as a safeguard against an unusually harsh or prolonged win-

FIGURE 3.6. Hay gathering is common in most species of pikas, such as the American pika (*Ochotona princeps*) living in the Rocky Mountains. It usually creates the hay piles in talus deposits, and these piles are crucial in providing food during the long winter period. *Source:* Gene H. Putney.

ter (Conner 1983; Dearing 1997). Size and placement of hay piles vary greatly among species. *O. princeps* establishes comparatively large hay piles, weighing up to 6 kg (13.2 lb), and up to 30 plant species may be found in one hay pile (Beidleman and Weber 1958). Daurian pikas (*O. dauurica*) inhabiting the steppes of northern Manchuria establish hay piles weighing 1 to 2.5 kg (2.2–5.5 lb) (VanderWall 1990). Pallas's pikas (*O. pallasi*) make hay piles measuring up to 100 cm (3.2 ft) in height, placed on the ground over burrow entrances. These pikas carry pebbles, up to 5 cm (almost 2 inches) in diameter, in their mouth and place them near the burrow entrance to prevent scattering of the hay by the wind.

Cheek Pouches

Among rodents, members of at least four families (hamsters, pocket gophers, pocket mice and kangaroo rats, and squirrels) have either internal or external cheek pouches that open near the angle of the mouth. External cheek pouches can be everted for cleaning. In many species, the mouth can close behind the incisors, and the animal may have either internal or external cheek pouches for transporting food. Cheek pouches are well adapted for carrying food; an early biologist discovered a total of 32 beechnuts in the cheek pouches of an eastern chipmunk (*T. striatus*) (E. G. Allen 1938). Heteromyids are well suited for granivory, with specialized fur-lined cheek pouches for the collection and transport of seeds and scratch-digging behavior for caching seeds over extended periods (Morton 1980; Morgan and Price 1992; Randall 1993). For example, kangaroo rats (*Dipodomys*), kangaroo mice (*Microdipodops*), and pocket mice (*Perognathus*) of North American deserts are primarily granivorous. Seeds are also the mainstay for tropical and subtropical species of heteromyids (*Heteromys* and *Liomys*) that harvest and cache fruits, nuts, and seeds from shrubs and trees. They collect large quantities of seeds and store them underground either in larders inside their burrows or scatter-hoarded in small buried caches outside the burrow. They use their large cheek pouches to collect many seeds in single foraging bouts. Seeds used by heteromyids are primarily from grasses and forbs (herbs other than grasses) and are quite small, usually less than 3 mm (0.1 inch) long and weighing less than 25 mg (0.0009 oz). Kangaroo rats collect most of their seeds directly from plants by clipping fruiting stalks and removing seeds from felled seed heads or by plucking seeds from fruit located close to the ground. Members of this family may also collect seeds, primarily in clumps, from the surface of the soil or strain them from the soil. Cached seeds are located by olfactory cues coupled with memory. Merriam's kangaroo rats (*D. merriami*) and pocket mice (*Perognathus amplus*) may locate seeds below the soil surface by detecting concentrated odors characteristic of buried seeds.

Six genera and approximately 40 species of pocket gophers of the family Geomyidae are restricted to North and Central America, from southern Canada through Mexico to extreme northwestern Colombia. They live in a variety of habitats that have soil conducive to burrowing. Geomyids are generally small, with a total length of 12 to 22.5 cm (4.7–8.9 inches), and weigh 45 to 400 g (11–32 oz) (fig. 3.7A). Males are always larger than females. Morphology reflects their fossorial mode of existence. Visual and auditory acuity is reduced in favor of enhanced tactile and olfactory sensitivity. Pocket gophers have thickset, chunky bodies with small eyes and **pinnae**. There are claws on the forefeet and hind feet for digging; those on the forefeet are larger. The incisors extend anteriorly, so the lips can be

FIGURE 3.7. (A) The plains pocket gopher (*Geomys bursarius*). (B) These gophers commonly establish "eskers" measuring up to several meters in length. Eskers are a common feature of the Colorado montane environment following snowmelt. *Sources:* (A) G. C. Hickman, ASM image library; (B) D. M. Armstrong.

closed behind them to keep soil out of the mouth when the animal is digging. The pectoral girdle is highly developed for digging, with a short, powerful humerus and a keeled sternum for enhanced muscle attachment, similar to that in the three families of moles. The pelvic girdle is small and relatively undeveloped. Gophers feed on **subterranean** parts of plants, primarily roots and tubers, although occasionally they leave their tunnels and forage aboveground. Food is carried back to storage chambers in large, externally opening, furred cheek pouches that extend from the mouth to the shoulder. Their underground tunnels may be quite extensive and include chambers for shelter, nesting, food storage, and fecal deposits. Tunnels are most easily detected by a series of aboveground mounds. Pocket gopher mounds are differentiated from those of moles by having a fan rather than a conical shape and by an off-center rather than central entrance hole.

Burrow systems of northern pocket gophers (*Thomomys talpoides*) may reach more than 150 m (164 yards) in length, usually located 10 to 45 cm (3.9–17.5 inches) below the surface. The side tunnels and chambers are commonly filled with food or feces or may be used for nesting. Northern pocket gophers are active during winter months and will subsist on roots, tubers, and bulbs. They may forage aboveground in snow tunnels (on the surface of the soil or within the snowpack), because the belowground tunnels are commonly saturated with meltwater. Their tunnels in the snow are packed with soil brought up from belowground. Following snowmelt in spring, these sinuous winter casts (sometimes called "gopher garlands" or "eskers"), measuring 5 to 10 cm (2–4 inches) in diameter and several meters in length, are a conspicuous feature of the montane landscape (Armstrong 2008: 117) (fig. 3.7B). Although gophers may damage croplands, gardens, and seedlings, they also aerate soil, increase water penetration into the soil, promote early successional plants (the first plants to colonize a habitat after disturbance), and enhance plants' diversity and community structure (Sherrod, Seastedt, and Walker 2005).

Other species well designed for digging are the tuco-tucos of South America (family Ctenomyidae) (fig. 3.8). The family contains more than 70 species of tuco-tucos in the single genus *Ctenomys*. These mammals are distributed from central South America south to Tierra del Fuego, living on grassy plains typified by sandy soils. Tuco-tucos range in size from the Los Talas tuco-tuco (*C. talarum*) of east-central Argentina, weighing about 100 g (3.5 oz), to the Tucumán tuco-tuco (*C. tucumanus*) of northwestern Argentina, weighing up to 700 g (1.5 lb). These fossorial rodents exhibit evolutionary **convergence** in habits and appearance with North American pocket gophers (Geomyidae). Like pocket gophers, tuco-tucos have broad, thick, prominent incisors, small eyes, a short, thick pelage, and enlarged claws. Unlike the geomyids, however, they lack external cheek pouches. The hind feet bear strong, bristled fringes. The tuco-tucos' scientific name, *Ctenomys*, means "comb-toothed" and is derived from the comblike nature of bristles around the soles of the hind feet and the toes, which are used to

FIGURE 3.8. The fossorial tuco-tuco (*Ctenomys boliviensis*) (family Ctenomyidae) from Santa Cruz Department, Bolivia. *C. boliviensis* is found in central Bolivia, southwest Brazil, western Paraguay, and Argentina. *Source:* Joseph Cook.

groom soil from the fur following a bout of foraging. The ctenomyids' burrow systems can be very extensive, and population densities may reach 200 individuals/ha. The Los Talas tuco-tuco is reported to "pack the earth of its tunnel walls firmly by standing on its forefeet, walking the hind feed up the wall, urinating, and then stamping the moistened soil into position as it lowers the hind feet" (Nowak 1999: 1680). Most digging takes place during daylight hours—tuco-tucos rarely emerge to forage until after dark. Tuco-tucos are quite vocal, and the "tloc-tloc" alarm call, which gave rise to the common name, can be heard when the animals are underground. Ctenomyids feed mainly on roots, stems, and grasses, which are probably pulled into burrows from belowground. Plants aboveground are secured by reaching from the surface entrance of the burrow (Puig et al. 1999; Justo, De Santis, and Kin 2003). The eyes of tuco-tucos are almost level with the top of the head, which is adaptive for surveying the horizon from their shelter without exposing themselves to predators. Reingestion of fecal pellets in Pearson's tuco-tuco (*C. pearsoni*) occurs while the animals are resting or between feeding bouts (Altuna, Bacigalupe, and Corte 1998).

Joe Cook of the University of New Mexico and Enrique Lessa of the Universidad de la Republica in Montevideo, Uruguay, are studying the evolution of tuco-tucos. Their findings indicate that tuco-tucos are a good example of explosive mammalian radiation over the past 2 million years, and more than 70 species are now recognized. Tuco-tucos have the widest range of chromosomal variation known for a mammalian genus, with diploid numbers ranging from 10 to 70, and they also show wide variation in sociality. Aspects of their fascinating physiology are only beginning to be uncovered, but they seem well adapted for the hypoxic (low-oxygen) environment of burrows and are found in habitats from sea level to dizzying heights of more than 4,572 m (15,000 ft) in the Andes Mountains (Cook, Anderson, and Yates 1990; Lessa and Cook 1998). Eileen Lacey of the University of California, Berkeley, is studying the social behavior and ecology of tuco-tucos. Most species of tuco-tucos are thought to be solitary, meaning that each adult lives alone in its own burrow system. At least two species, however, have abandoned this solitary lifestyle to live in groups composed of multiple adults and young, all of whom share a burrow system and nest. These more gregarious species are the colonial tuco-tucos (*C. sociabilis*), found in southwestern Argentina, and the Peruvian tuco-tucos (*C. peruanus*), found in the Altiplano, south of Lake Titicaca. Understanding why the behavior of these species differs is a challenging task and includes analyses of their ecology, physiology, demography, and neurobiology. Given how many species of tuco-tucos have yet to be studied, it seems likely that other examples of group living among these secretive animals remain to be discovered.

Frugivores

Animals that exhibit adaptations to a diet of fruits, the reproductive parts of plants, are referred to as **frugivorous** ("fruit-eating"). Mammals from several orders are known to specialize in the consumption of fruit. Practitioners include many species of flying foxes and leaf-nosed bats, as well as the marsupial phalangers and even the treeshrews. Many primates are frugivorous (Emmons 1991; Feldhamer et al. 2007). Because fruits may have a hard outer covering, the teeth of some frugivores are adapted for piercing and crushing. Mammals that subsist on softer fruits typically have a reduced number of cheekteeth with a low-crowned (**bunodont**) occlusal pattern.

Piercing Teeth

Many species of bats depend on plants for food. The frugivorous bats are distributed in two families: the Pteropodidae of the Old World tropics and Phyllostomidae of the New World tropics (Patterson, Willig, and Stevens 2003). Bats play an essential role in the pollination of flowers—a form of pollination known as chiropterogamy—and dispersal of seeds. Some seeds have higher rates of germination after passing through the gut of a mammal, and this method of seed and fruit dispersal by bats is termed chiropterochory. Frugivorous bats include either principal or partial pollinators and pollen dispersers for close to 130 genera of tropical and subtropical plants. Bats are particularly important to plant species that blossom only at night, such as avocados (*Persea*), balsa (*Ochroma lagopus*), durian (*Durio zibethinus*), *Eucalyptus,* figs (*Ficus*), guava (*Psidium guajava*), kapok (*Ceiba pentandra*), mango (*Mangifera indica*), papaya (*Carica papaya*), and wild banana (*Musa paradisiaca*).

Megachiropterans (Old World flying foxes and fruit bats) are restricted to the tropical forests of Africa, Asia, and the Australian region, where succulent fruits are plentiful. Body size varies considerably among members of this family. The smaller species, such as the long-tongued fruit bat (*Macroglossus minimus*) (fig. 3.9A), pygmy fruit bat (*Aethalops alecto*), and spotted-winged fruit bat (*Balionycteris maculata*), weigh only about 15 to 20 g (0.5–0.7 oz); the largest species in the genus *Pteropus* weigh up to 1.2 kg (2.6 lb)—a difference of two orders of magnitude. Megachiropterans feed on fruits of at least 145 plant genera in 50 families. Most bat species locate fruit by smell. Old World fruit bats have comparatively few teeth, and the lower molars are reduced in number and have large, flat grinding surfaces (fig. 3.9B). The canines are the principal piercing teeth. Megachiropterans can bite into fruit while hovering, or they may hang onto a branch with one foot, biting into the fruit as they press it to their chest with the other foot. If the fruit is small, they may carry it with them to a

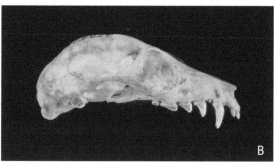

FIGURE 3.9. (A) A representative frugivorous bat, the long-tongued fruit bat (*Macroglossus minimus*). (B) Cranium, showing teeth, of the little golden-mantled flying fox (*Pteropus pumilus*). *Sources:* (A) Fritz Geiser; (B) University of Michigan, Animal Diversity Web, http://animaldiversity.org, courtesy of Phillip Myers.

branch and hang head downward while they consume it. Rather than biting off and swallowing mouthfuls of fruit, most megachiropterans crush pulp from soft fruits and extract only the juice and reject the pulp and seeds, or they may digest the pulp and excrete intact seeds. Because this activity often occurs at some distance from the harvesting site, seeds are dispersed over great distances. Several species of Old World fruit bats (*Rousettus aegyptiacus, Epomophorus wahlbergi,* and *Eidolon helvum*) are principal agents of dispersal of the baobab (*Adansonia digitata*), an important tree in the African savannah.

Although the New World leaf-nosed bats may be viewed as **generalists,** more than one-half of the species consume some fruit. Most fruit-eating phyllostomids are in the subfamilies Carolliinae or Stenodermatinae of the neotropics or Brachyphyllinae of the Antilles. Members of the subfamily Carolliinae have reduced molars and consume ripe, soft fruits such as bananas and figs. In contrast, members of the subfamilies Stenodermatinae and Brachyphyllinae have more robust molars. Large species such as the Jamaican fruit-eating bats (*Artibeus jamaicensis*), averaging about 46 g (1.6 oz) in body mass, are generalist frugivores widespread in the neotropics (fig. 3.10A, B). Their teeth are well adapted to crushing fruit, aiding in their consumption of up to 92 taxa of plants. In moist tropical forests of Barro Colorado Island in Panama, *Artibeus* relies on two abundant species of figs that produce very heavy crops of fruit. For example, *Ficus insipida* is known to produce 40,000 fruits during one week once or twice per year. Like many of the megachiropterans, *Artibeus* (a microchiropteran) bites chunks from a fruit and crushes the pulp for its juice with its broad, flat posterior teeth (fig. 3.10C). These bats eat their own body mass in fruit each night, and food passes through their alimentary tract rapidly, in about 15 to 20 minutes. *Artibeus* is important in the dispersal of seeds of tropical fruits (Flemming 1982). Terry Vaughan reports that studying the feeding ecology of *Artibeus* can be a bit tenuous. While he and his colleagues were camping beneath a fig tree in central Sinaloa, Mexico, the foraging activity of several species of *Artibeus* prompted the camper-researchers to disperse rather rapidly from their vantage point. "The activities of dozens of *A. lituratus, A. hirsutus,* and *A. jamaicensis* caused a nearly continuous rain of fruit and bat excrement throughout much of the night, and with sunrise came herds of aggressive local pigs to gather the night's fallout of figs" (Vaughan, Ryan, and Czaplewski 2000: 167). Such an incident attests to the efficient foraging ability of *Artibeus!*

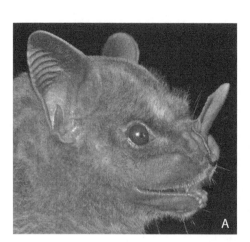

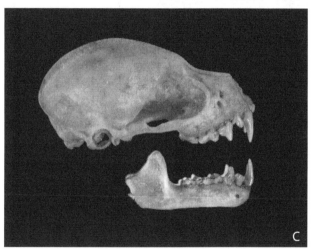

FIGURE 3.10. (A) The head and (B) the body in flight of a Jamaican fruit-eating bat (*Artibeus jamaicensis*). (C) Skull of the same species, showing teeth that are well adapted to crushing fruit. *Sources:* (A, B) Heather A. York; (C) University of Michigan, Animal Diversity Web, http://animaldiversity.org, courtesy of Phillip Myers.

Prehensile Tail and Protrusile Tongue

The **prehensile** tail, stereoscopic vision, and frugivorous habits of kinkajous were once sufficient evidence to categorize them as primates. So convincing was the evidence that eighteenth-century taxonomists assigned these **arbo-**

real members of the order Carnivora as *Lemur flavus,* in the order Primates (Schreber 1774). Studies of their anatomy and genetics have shown that kinkajous belong with the raccoons, coatis, olingos, ringtails, and cacomistles as members of the New World family Procyonidae. Kinkajous (*Potos flavus*) (fig. 3.11) inhabit tropical forests of Central and South America. Weighing 2 to 4.6 kg (4.4–10.12 lb), they have a thick and woolly, honey-brown pelage; they are very similar in appearance and diet to olingos (*Bassaricyon*). The two may forage side by side, with the olingo preferring insects, small mammals, and birds, and the kinkajou focusing on fruits. Known by locals as *mono de la noche,* or "monkey of the night," the kinkajou is the most arboreal of all procyonids. Aided by their long, fully prehensile tail, nimble clawed fingers, and fully reversible hind feet, kinkajous commonly hang upside down during feeding. These solitary **nocturnal** procyonids are mainly frugivorous, with a diet consisting of 90% fruit and 10% nectar and leaves. Fruits are usually consumed when ripe, with figs being a staple in the kinkajou diet. Preference for figs may be due to their high nutrient content compared with other tropical fruiting species. Kinkajous have been observed using their long, protrusile tongue to probe flowers

FIGURE 3.11.   The kinkajou (*Potos flavus*), a frugivorous member of the raccoon family, resides in tropical forests of Central and South America. *Source:* Rexford D. Lord.

in search of nectar or for obtaining honey from beehives. Other food items include leaves, buds, flowers, insects, and small vertebrates. With the exception of the fruit bats, few mammals are as committed to the frugivorous way of life as are the kinkajous.

Nectarivores

Insects and hummingbirds are not the only animals that feed on flowers. Some mammals are also exquisitely adapted to feed on nectar and, in the process, transfer pollen. **Nectarivorous** ("nectar-eating") mammals are represented by about six genera of bats (both megachiropterans and microchiropterans) and the marsupial honey possums. Their skulls are characterized by elongated snouts; small, weak teeth (bats); reduced numbers of teeth (honey possums); and poorly developed jaw musculature.

Brush-Tipped Tongue

The tongue of nectarivores is long, slender, and protrusile and typically has a brush tip consisting of many rows of hairlike papillae pointing toward the throat. Bats of the subfamily Glossophaginae (family Phyllostomidae) are the most specialized mammalian nectarivores. The Mexican long-tongued bat (*Choeronycteris mexicana*) of Central America, Mexico, and the southwestern United States has a distinctively long and slender skull equipped with a long **rostrum** well adapted to feed on fruits, pollen, nectar, and insects (Arroyo-Cabrales, Hollander, and Knox Jones 1987; Elbroch 2006). Flowers of the *Agave* and the saguaro cactus (*Carnegia gigantea*) are staples in the diet of the Mexican long-nosed bat (*Leptonycteris nivalis*), a bat (weighing 20–25 g, or 0.7–0.9 oz) with high energy demands. Long-nosed bats commonly hover while feeding and will use their long snout and flexible tongue to probe into the corolla of *Agave*, for example, to extract nectar. The bat licks nectar with its tongue, which can be extended up to 76 mm (almost 3 inches). Although its goal is to secure nectar, *L. nivalis* will invariably become coated with pollen while feeding. The animal then unavoidably transports this pollen to the next flower, thereby ensuring fertilization. Mexican long-nosed bats are agile flyers and form foraging flocks of at least 25 bats that feed at successive plants. While feeding, they circle the plants and take turns feeding on the flowers. Flocks show a cohesiveness typified by minimal antagonistic behavior. Flocking seems to confer the adaptive advantage of an increased foraging efficiency, which is critical for minimizing energy expenditure. The Mexican long-tongued bat (*C. mexicana*) is another phyllostomid regarded as an obligate pollen-feeder. Analysis of the stomach contents of bats from Central America showed a majority of pollen grains from pitahaya (*Lemaireocereus*), cazahuate (*Ipomoea*), *Ceiba, Agave,* and garambulla (*Myrtillocactus*) (Arroyo-Cabrales, Hollander, and Knox Jones 1987).

One of the most highly specialized nectar-feeding bats is a recently described glossophagine, the tube-lipped nectar bat (*Anoura fistulata*) (fig. 3.12A, B), an inhabitant of the cloud forests of the Andes Mountains of Ecuador (Muchhala and Jarrin-V 2002; Muchhala, Mena, and Albuja 2005). The tube-lipped nectar bat coinhabits the cloud forests with two other glossophagines, *A. caudifer* and *A. geoffroyi*. Nathan Muchhala of the University of Toronto found that this unique bat has a tongue that extends 84.9 mm (3.3 inches)—one-and-a half times its body length and twice as long as the tongue of its congeners (bats of the same genus). *A. fistulata* has the longest tongue ever reported in a mammal, surpassed only by that of the chameleon among vertebrates. The morphology of its tongue is indeed unique (fig. 3.12B–D). In other nectarivorous bats, the base of the tongue attaches to the base of the oral cavity. However, the tongue of the nectar bat passes back through the neck, attaching in the thoracic cavity, where its distal portion is surrounded by a sleeve of tissue (the glossal tube) that parallels the ventral surface, with the base inserting between the heart and sternum. *A. fistulata* is the exclusive pollinator of the plant *Centropogon nigricans,* which has corollas of matching length—80 to 90 mm. Both *A. fistulata* and pangolins (order Pholidota, family Manidae) have glossal tubes to accommodate attachment of their long, protrusile tongues (Chan 1995). In a fascinating example of coevolution, as the flower of *C. nigricans* became longer over millennia, the bat's tongue elongated to reach the nectar stored at the base of the ring of petals. As Nathan states, "It is like a cat being able to lap milk from two feet away" (Bat Conservation International 2006). Tube-lipped nectar bats and pangolins represent another excellent example of convergent evolution—the evolution of similar morphologies (long, protrusile tongues) by distantly related lineages as adaptive solutions to similar ecological pressures—in this case, availability of ant- and pollen-feeding niches.

Parallel adaptations for nectarivory are well known in megachiropter-

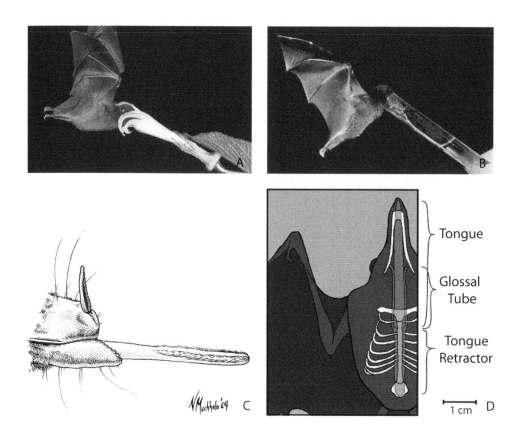

Tongue

Glossal
Tube

Tongue
Retractor

1 cm   D

FIGURE 3.12.   The tube-lipped nectar bat (*Anoura fistulata*). (A). *A. fistulata* pollinat-ing the specialized flower of *Centropogon nigricans;* only *A. fistulata* can reach the nectar at the base of the long corolla. (B) *A. fistulata* feeding from a test tube filled with sugar water; note that its tongue (pink) can extend to 150% of its body length. (C) Lateral view of the nose leaf, lip, and partially extended tongue, showing prox-imally facing papillae. (D) Ventral view showing the tongue (pink), glossal tube and tongue retractor muscle (blue), and skeletal elements (white*). Sources:* N. Muchhala (*Nature* 444:701, 2006); N. Muchhala et al. (*Journal of Mammalogy* 86:457–461, 2005).

ans such as the long-tongued fruit bats (*Macroglossus*) and blossom bats (*Syconycteris*) (figs. 3.9A and 3.13). Southern blossom bats (*S. australis*) of eastern Australia feed on nectar and pollen of a variety of rainforest plants, such as cauliflorous (*Syzigium*), *Banksia, Melaleuca, Callistemon,* and some *Eucalyptus.* Blossom bats locate nectar and pollen with their large eyes and keen sense of smell. *Syconycteris* typically lands on an inflorescence and gathers pollen and nectar by use of its long snout and brush-tipped tongue—

it does not hover while feeding, as do many other nectarivores. These bats are unique in that they do not consume pollen directly from the flower. Their body hairs are covered with small, scalelike projections in which pollen lodges. This pollen is "consumed" while the animal grooms its fur and wings after a foraging episode. As with other nectarivores and frugivores, pollen rapidly passes through the gut and appears in the feces 45 minutes after ingestion. The majority of protein in the diet of blossom bats is provided by the pollen, and sugars in the nectar help to meet energy demands (Law 1992, 1993).

A unique marsupial species is the premier **terrestrial** nectarivore: the honey possum, or noolbenger (*Tarsipes rostratus*), the only species in the marsupial family Tarsipedidae (fig. 3.14A). These mouse-sized possums are found only on the sand-plain heaths of the southwestern tip of Western Australia, where there is a rich diversity of nectar-producing plants (families Proteaceae and Myrtaceae) that provide food throughout the year (Renfree, Russell, and Wooller 1984). Interestingly, the honey possum is the only

FIGURE 3.13. The frugivorous common blossom-bat (*Syconycteris australis*) of Australia. *Source:* Fritz Geiser.

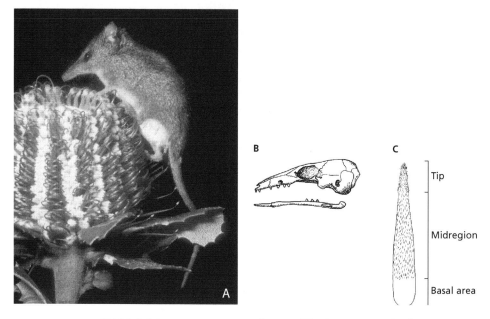

FIGURE 3.14. (A) Male honey possum, or noolbenger (*Tarsipes rostratus*), of western Australia, on a *Banksia* inflorescence. (B) Except for the procumbent lower incisors, the honey possum has small, degenerate teeth, reflecting a soft diet. (C) The long, extensible, brush-tipped tongue is used to procure nectar and pollen, a staple in its diet. *Sources:* (A) P. A. Woolley and D. Walsh; (B, C) Feldhamer et al. 2007.

terrestrial mammal that is *entirely* nectarivorous; it fills an ecological niche in Australia similar to that occupied by hummingbirds in the New World. *Tarsipes* is one of the smallest of all possums, almost as tiny as the diminutive dasyurids, *Ningaui* and *Planigale*. Adult male honey possums weigh 7 to 9 g (0.2–0.3 oz); females are always dominant to males and weigh slightly more, 10 to 12 g (0.3–0.4 oz).

Honey possums subsist exclusively on the nectar and pollen of several species of flowering shrubs, especially *Banksia*. Unlike most mammals, this mouse-sized marsupial does not climb with the aid of claws but has digits that are expanded at the tip with short, nail-like structures adaptive in gripping branches. This morphology is similar to that in primates, and thus the generic name (*Tarsipes*) refers to a supposed resemblance of the honey possum's feet to those of the tarsier, a primitive primate of islands of Southeast Asia. A pair of lower incisors are the only well-developed teeth (fig. 3.14B). The teeth number 20 at most and are reduced to small pegs, reflecting the possum's soft diet. Possums are active, nimble climbers, locating food by smell; they insert their long snout into a flower and use their protrusile tongue (reaching some 25 mm, or almost 1 inch, beyond the nose)—which has bristles at the tip, stiffened with a keratinous keel—to lick pollen from protruding anthers. Tongue and palate work in harmony to collect pollen. The pollen is scraped from the upper surface of the tongue by transverse combs on the roof of the mouth and transported to the stomach. The stomach is small, housing an accessory pouch analogous to the crop of birds or the cheek pouches of a chipmunk. It acts as a temporary storage compartment for the nectar collected during the night; the intestine is short, and there is no cecum. The stomach pouch may also serve as a site for sugar storage. Nectar contains about 25% sugar and fills the water and energy needs of honey possums; pollen contains 20% protein and 37% carbohydrate and provides all of the possum's protein and energy needs (Wooller et al. 1981; F. J. Bradshaw and Bradshaw 2001; Tyndale-Biscoe 2005). Honey possums are reported to have an average **metabolic rate** twice that of any marsupial and 158% of the average rate for **placental** mammals (Withers, Richardson, and Wooller 1990). To meet their energy needs, honey possums must feed on at least 14 to 17 *Banksia* inflorescences in an area of 0.7 ha (1.7 acres) per day; because of the wealth of pollen and nectar in the sand-plain heaths, a honey possum can meet its entire energy needs in an area as small as 104 km^2 (40 square miles) (Wooller et al 1981; Wooller, Russell, and Renfree 1984).

A combination of maritime climate, mild winter, and diversity of woody shrubs flowering throughout the year optimizes the survival of honey possums in the southwestern tip of Australia. Nectar-feeding birds such as New Holland honeyeaters (*Phylidonyris novaehollandiae*) and western spinebills (*Acanthorhynchus superciliosus*) are potential competitors for food in the form of pollen, including the pollen of *Banksia*. However, honey possums feed extensively on prostrate and bushy species of *Banksia* (*B. nutans*) that are inaccessible to honeyeater birds. It seems that these low-lying species of *Banksia* rely heavily on honey possums for pollination and may have coevolved with *Tarsipes* in the coastal heathlands (Wooller, Russell, and Renfree 1984).

Another Australian nectarivore, the feather-tailed glider (*Acrobates pygmaeus*) (fig. 3.15), is the smallest gliding mammal in the world (Goldingay and Scheibe 2000). A denizen of tall forests and woodlands of eastern Australia, *A. pygmaeus* shows some striking adaptations to its environ-

FIGURE 3.15.   The feather-tailed glider (*Acrobates pygmaeus*) resides in forests and woodlands of eastern Australia. It is the smallest gliding mammal in the world and an obligatory nectarivore. *Source:* Fritz Geiser.

ment. Like the New World flying squirrels, this small, agile, pygmy acrobat is a top-notch glider and can achieve glides of some 20 m (66 ft) between trees. On the chewing end of things, it has "insectivore"-like teeth coupled with the brush-tipped tongue of a nectar-feeder. Its main foods are nectar, sap, manna, insects, and foliage. Leaping locomotion is enhanced by its forward-directed eyes adapted to nocturnal binocular vision. Like tree frogs and geckos, *Acrobates* has sharp claws for climbing and feet equipped with striated pads on each toe to enhance adhesion to the smooth bark of euca-lypts. Finally, its featherlike, partially prehensile tail assists with grasping branches. The tail, fringed with hairs along the sides, also augments the an-imal's aerodynamic activities. Due to their small size (less than the mass of a house mouse) and nocturnal habits, feather-tailed gliders are frequently overlooked by visitors to their habitat. *A. pygmaeus* makes small leaf nests in hollow trees and branches 15 m (49 ft) above the forest floor.

....

Animals that primarily consume plant exudates, such as resins, sap, or gums, are termed **gummivorous** ("gum-eating") (Bearder and Martin 1980; Nash 1986). This peculiar dietary specialization occurs in eight species of marmosets (*Callithrix* and *Cebuella*), bushbabies (*Galago senegalensis*), pottos (*Perodicticus*), and slow lorises (*Nycticebus coucang*) (Wiens, Zitzmann, and Hussein 2006); in four species of petaurid gliders (*Petaurus*); and in Leadbeater's possum (*Gymnobelideus leadbeateri*). All members of the family Cheirogaleidae (dwarf and mouse lemurs) feed on tree exudates; however, consumption of gum is only occasional for *Microcebus*, *Cheirogaleus*, and *Mirza*. Gums are a staple in the diet of *Phaner* and *Allocebus*.

A champion gummivore is the fork-marked mouse lemur (*Phaner furcifer*) of Madagascar (fig. 3.16). This little primate weighs 300 to 500 g (10.6–17.6 oz), and its diet consists of close to 90% gum exudates from the trunks and branches of trees. It has long hind limbs and relatively large hands and feet, with large digital pads. Mouse lemurs are equipped with clawlike fingernails and long anterior premolars that are caniniform; they use a "tooth comb" (formed by their **procumbent** lower incisors and canines) to scrape off gum released from the surface of a plant. During feeding, lemurs easily cling to the surface of the trunk by their needle-sharp claws. The fork-marked lemur consumes gum from orifices created by beetle larvae that live between the wood and bark of trees. The lemur's long, narrow tongue assists in obtaining gum, and a specialized species of symbiotic bacteria in the cecum assists in digesting this staple food. The gums are poor in nitrogen (i.e., in protein), so the lemurs also consume insects as a source of protein (Charles-Dominique and Petter 1980). The main trees used as sources of gum are *Terminalia* species. The fork-crowned lemur usually is nocturnal, foraging at heights of 8 to 10 m (26.2–32.8 ft) and often in the tops of trees (Hladik, Charles-Dominique, and Petter 1980). Lemurs are able to scrape off saps and gums exuded as a result of damage by wood-boring insects. Marmosets are thought to be the only primates that actually gouge holes to liberate plant juices.

The black-tufted marmoset (*Callithrix penicillata*), an inhabitant of rainforests of Brazil's central plateau, regularly consumes tree sap by nibbling the bark with its long lower incisors. The incisors have thickened

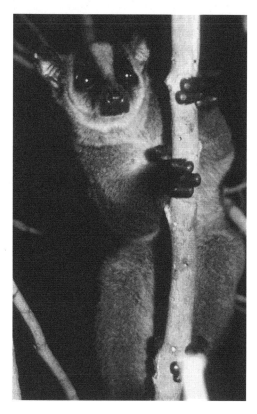

FIGURE 3.16. The fork-marked mouse lemur (*Phaner furcifer*) of Madagascar. This small primate's diet consists of close to 90% gum exudate obtained from the trunks and branches of trees. *Source:* Russell A. Mittermeier, Conservation International.

enamel on the outer surface and lack enamel on the inner surface, thus producing chisel-like instruments. Anchoring their upper incisors in the bark, marmosets use their lower incisors to gouge oval holes in the tree trunk. These holes may measure 2 to 3 cm (0.8–1.2 inches) across, and certain trees may be riddled with channels 10 to 15 cm in length (Tattersall 1982). As in the lemurs, the clawlike nails of marmosets are essential adaptations for clinging to vertical trunks while feeding on sap, gums, and resins. The pygmy marmoset (*Callithrix pygmaea*), the world's smallest true monkey, also feeds on tree sap and gum and insects, and occasionally on fruits (fig. 3.17A). These tiny monkeys bite numerous small (8–15 cm, or 3.1–5.8 inches) round holes in the bark of trees and vines and will return day after day to feed on the sap or gum that seeps into the wounds. Each group of marmosets feeds on a few such trees, some with hundreds of old and currently used feeding holes.

Plant and insect exudates make up the bulk of the diet of several members of gliding and striped possums (family Petauridae) of Australia, Tasmania, and New Guinea. One group of interesting "aussie gliders" is the lesser gliding possums (*Petaurus*). They resemble the North American flying squirrels (*Glaucomys*) in form by possessing a large gliding membrane

FIGURE 3.17. Selected gummivores. (A) The world's smallest monkey, the pygmy marmoset (*Callithrix pygmaea*) of the family Cebidae, an inhabitant of the rainforests of Brazil's central plateau. (B) Yellow-bellied glider (*Petaurus australis*) of Australia. (C) Bushbaby (*Galago*) of Africa. *Sources:* (A) Rexford D. Lord; (B) A.N.T. Photo library / NHPA/Photoshot; (C) Luboš Mráz.

(**patagium**); however, the tail of the Australian possums is furred all around and not as flattened as that of *Glaucomys*. All species inhabit wooded areas and are arboreal and nocturnal. Sugar gliders (*P. breviceps*) consume mostly nectar and arboreal arthropods; they also remove bark from certain species of *Eucalyptus* to gain access to the sugary sap. In doing so, they make characteristic V-shaped incisions in the bark that channel the sap to their mouth, which is strategically placed at the bottom of the V. Reportedly, sugar gliders may spend 90% of the time outside their dens in foraging—and 70% of that foraging time is dedicated to feeding on the nectar of *Eucalyptus* (Goldingay 1990). Understandably, *P. breviceps* is crucial in facilitating cross-pollination of *Eucalyptus* trees. Like southern flying squirrels, *G. volans*, sugar gliders huddle together in hollow trees, in groups of up to seven adult males and females and their young, to conserve energy during cold weather. Unlike flying squirrels, they may enter **torpor** when food supplies are low. The largest of all the petaurids are the yellow-bellied gliders (*P. australis*) (fig. 3.17B), weighing 500 to 700 g (17.6–24.7 oz). These are the true **specialists** in sap feeding. They use their powerful incisors to cut the bark of particular trees and feed for many hours of the night on flowing sap. Sap is the principal source of energy for yellow-bellied gliders, and they are very selective in the trees that they tap. They forage on sap from *Eucalyptus* by biting out small patches of bark on the trunk or main branches. After the flow of sap dries up, they move on to a new area. As a result, some trees become heavily scarred after several years of feeding (Goldingay and Kavangah 1991). In southern Australia, these gliders are rather catholic in their preferences, foraging on some 24 species of *Eucalyptus,* but in northern Queensland they exploit only one species, the red mahogany (*E. resinifera*).

Ross Goldingay and his colleagues at Southern Cross University in New South Wales have conducted a great deal of research on gliders. They recently published an interesting article on the **home range** of squirrel gliders (*Petaurus norfolcensis*)—a threatened species and one that inhabits an area much affected by human development (Sharpe and Goldingay 2007). This handsome marsupial glider, a relative of the sugar glider, weighs 180 to 300 g (6.3–10.6 oz) and lives in **sclerophyll** forests and woodlands of southeastern Australia. Like sugar gliders, this species forms communal groups in hollow trees. Nectar and pollen are dietary staples; when these are not available, squirrel gliders will switch to eucalypt sap and wattle gum. Goldingay and colleagues, using radiotelemetry techniques and several home-range estimators, concluded that the home ranges of squirrel gliders are strongly influenced by the distribution of key winter- and spring-flowering trees.

Leadbeater's possums (*G. leadbeateri*) are secretive, highly arboreal possums inhabiting mountain forests of the central highlands of Victoria, Australia. Like the mountain pygmy possums, the trail of Leadbeater's possum has been tough to track. Known from only five specimens found between 1867 and 1909, Leadbeater's was thought to be extinct—until its rediscovery in 1961. Since this time, refined search techniques have been very productive, expanding the known distribution to close to 200 localities throughout the central highlands of Victoria. This nongliding member of the family Petauridae does not seem to be hampered by a lack of gliding membrane; it readily leaps from tree to tree in meter-long jumps. Their rapid movements, nocturnal foraging, and shy disposition make these possums difficult to observe. Leadbeater's possums depend on large-diameter hollowed mountain ash (*Eucalyptus regnans*) for nesting sites, which accommodate eight or more communal nesters. Specialized apical toe pads permit a firm grip on the smooth bark of eucalypts. The diet of *G. leadbeateri* is very similar to that of other gliders. They glean insects from beneath the shedding bark of eucalypts and consume exudates, including the gums of silver and hickory wattles, manna produced at sites of insect damage, and honeydew secreted by sap-sucking insects. Their diet of plant exudates is augmented by arthropods, an important source of protein.

Bushbabies, also known as galagos, are members of the primate family Galagidae (fig. 3.17C). All members of this family live in Africa, inhabiting rainforests in West Africa and woodland savannah from Senegal to East Africa and south to southern Africa. All members of the family are arboreal. A key feature of this group is the mode of locomotion, which involves leaping and bounding from branch to branch and between tree trunks. Two adaptations—well-developed hind limbs (longer than their forelimbs) and long, bushy tails used for balance—aid in this form of locomotion. Senegal bushbabies (*Galago senegalensis*) hold jumping records among galagos, with leaps of 5 m (16 ft) being routine. Galagids range in size from the diminutive Prince Demidoff's bushbaby (*G. demidoff*), which has a head and body length of 120 mm (4.7 inches) and mass of only 60 g (2.1 oz), to the much larger Brown greater galago (*Otolemur crassicaudatus*), with a head and body length of 320 mm (12.6 inches) and mass of 1.2 kg (2.6 lb). All bushbabies are **pentadactyl,** with the second toe modified as a toilet claw that is used for grooming. Their diet includes fruits, gums, nectar, insects, eggs, and various small prey, including birds, rodents, and bats. As discussed below, some bushbabies subsist primarily on gums. Bushbabies have a soft, woolly fur, ranging in color from gray to brown and russet brown, with lighter undersides. Field studies, limited primarily to thick-tailed galagos and Zanzibar bushbabies (*Galago zanzibaricus*), show that galagids live in groups of up to seven to nine individuals and build well-concealed nests in trees or use cavities in hollow trees. Single births or twins occur for all species studied. Mothers nurse their young with two to three pairs of **mammae** (mammary glands). Early in their infants' development, mothers carry their young with their canines, gripping them by the scruff of the neck. Bushbabies communicate by facial expressions, urine marking, and vocalizations, notably the crylike sound made by young bushbabies, from which the group gets its common name. They are nocturnal and have large eyes that allow them to see well at night. They typically feed alone. Like other strepsirhine primates (suborder Strepsirhini), bushbabies have forward-pointing (**procumbent**) incisors and incisorlike canines that form a tooth-comb-like arrangement that is important in grooming and feeding. Bushbabies are unique in possessing a cartilaginous brushlike structure (the sublingua) on the underside of their tongue. The points of the sublingua act a bit like toothpicks, dislodging particles of insects or gum from between the comblike teeth.

Gummivory is common among the galagos; this feeding strategy is perfected in southern needle-clawed bushbabies (*Euoticus elegantulus*). A permanent diet of gum was probably a staple in the diet of *Euoticus* at a very early stage in its evolution, because its alimentary canal shows specializations not present in other galagos. Gums are an essential food source for lorisiforms (bushbabies, lorises, and pottos); however, gums contain long-chain sugars that can only be digested through bacterial fermentation in the gut. Lorisiforms have an enlarged cecum to deal with this function—the cecum of the southern needle-clawed bushbaby is five times larger than expected for its body mass!

The geographic range of needle-clawed bushbabies closely adheres to the distribution of their staple food, lianes (lianas) and gum-producing trees of Africa: 80% of their diet consists of resins and gums produced by deciduous trees (*Newtonia*) and legumes such as *Albizia*. The sharp nails and exceptionally broad hands of *E. elegantulus* allow it to surmount and forage on large tree trunks that are inaccessible to most galagos. When foraging for gum, it uses a regular pathway of trees, stopping at each one every night. To gain sufficient fresh gum, these galagos must frequent numerous trees and shrubs and establish many oozing sites that they visit each night. They may stop at 500 to 1,000 gum-feeding locations in a single night. Due to its delicate incisors and premolars, *Euoticus* is unable to establish excavation wounds that are deep enough to keep gum flowing. However, it has overcome this obstacle by recruiting some help—some of which comes in the form of very distant cousins. In western Africa, a principal source of gum droplets results from bark wounding by scaly-tailed squirrels (family Anomaluridae), coupled with help from cicadas that produce tiny bore holes. Interestingly, cicada activity in **mesic** forests of central West Africa may be massive and continuous throughout the year, yet needle-clawed bushbabies rarely consume the cicadas. The limited distribution of *Euoticus* in humid forests of western Africa seems to be governed by the presence of these bark-wounding animals—scaly-tailed squirrels and cicadas (Kingdon 1997b; Macdonald 2006).

Folivores

Animals that consume primarily leaves are termed **folivorous** ("leaf eating"). The alimentary canal of such herbivores is typically adapted for feeding on cellulose-rich herbs and grasses for which mammals lack digestive enzymes. Although the specialized teeth of herbivores effectively shred and grind the cell walls of plant tissues and release their contents, only certain enzymes can digest cellulose. Mammals do not produce these **cellulolytic** (cellulose-splitting) enzymes, so they rely on symbiotic microorganisms residing in their alimentary canal. These microorganisms break down and metabolize cellulose and release fatty acids and sugars that can be absorbed and used by the mammalian host. Like the hoofed mammals, rodents and lagomorphs cannot produce the enzyme cellulase, so the fermentation of fibrous forage is carried out with the aid of bacteria and protozoa. Young rodents and lagomorphs become inoculated with the appropriate anaerobic protozoans and bacteria by eating maternal feces, whereas young ungulates commonly acquire their microorganisms by consuming soil.

Foregut versus Hindgut Fermentation

Rumination, also called **foregut fermentation,** is carried out by artiodactyls (even-toed hoofed mammals) such as camelids, giraffids, hippopotamuses, antilocaprids, cervids (deer), and bovids, as well as by kangaroos, sloths, and colobus monkeys (Freudenberger, Wallis, and Hume 1989; Alexander 1993; Stevens and Hume 1996). In foregut fermenters, plants (such as grasses) with hard-to-digest cell walls are dealt with by a complex, multi-chambered stomach containing cellulose-digesting microorganisms. After food is procured by cropping or grazing, it immediately passes to the first and largest chamber of the network, the **rumen** (see Feldhamer et al. 2007: fig. 7.9). Here, the food is moistened and kneaded, which mixes it thoroughly with microorganisms that ferment the food. Large particles of food float on top of the rumen fluid and pass to the second chamber, the **reticulum**, a blind-ended sac with honeycomb partitions in its walls. The reticulum is where a softened mass called the "cud" is formed. Fermentation occurs in both the rumen and reticulum, and both chambers absorb the main products of fermentation, short-chain fatty acids. When the animal is at rest, the cud is regurgitated, allowing the animal to "chew its cud," or **ruminate**. At this time, the mass is further broken down by a potent enzyme, **salivary amylase**, produced by the ruminant animal. The food is then swallowed a second time and enters the third chamber, the **omasum**, where muscular walls knead it further. The fourth chamber, the **abomasum**, is the true stomach. Here, digestive enzymes that kill any remaining microorganisms are secreted, and protein digestion is completed. Digested material then passes into the small intestine, where the products of microbial digestion and acid digestion are absorbed. Some additional fermentation and absorption occurs in the cecum.

Like the perissodactyls (odd-toed hoofed mammals), rodents and lagomorphs do not ruminate, and **hindgut fermentation** occurs in the colon and cecum. The only rodents that lack a cecum are dormice (family Gliridae), indicating that their diet includes little cellulose. Compared with ruminants, the stomach of rodents is simple, but it does have up to three chambers. The small intestine is comparatively short, and the hindgut (colon and cecum) is complex, with the cecum having many spiral folds, recesses, and saclike expansions. In the order Rodentia, variation in the morphology of digestive systems is correlated with diet. For example, members of the family Sciuridae (squirrels, chipmunks, and marmots), which feed on a variety of seeds, nuts, fruits, and herbs, have a much simpler digestive system than grass-eating voles and lemmings.

Plant-eaters typically have a long intestine with either a simple stomach (nonruminant herbivores) or one that has internal folds and, as noted above, is divided into several functionally different chambers. Besides rodents and lagomorphs, hindgut fermentation is characteristic of horses, tapirs, rhinoceroses, howler monkeys, elephants, hyraxes, and some arboreal marsupials (Alexander 1993). Hindgut fermenters masticate food as they eat, initiating digestion with salivary enzymes. Digestion continues by enzymatic activity in the simple stomach, and food then moves rapidly into the small intestine as new food is eaten. Unlike ruminant artiodactyls, hindgut fermenters do not regurgitate their food. Nutrients are absorbed in the small intestine. Finely ground particles of food pass from the small intestine into the cecum, and larger food particles move through the large intestine and are passed as feces. Among the hindgut fermenters, the colon acts as the principal fermentation chamber for the larger species, while the cecum fulfills this function in the smaller species (Hume 1989).

The two kinds of fermentation process in herbivores have clear advantages and disadvantages (Montgomery 1978; Dawson 1995). Foregut fermentation tends to be very efficient because microorganisms begin to break down the plant material before it reaches the small intestine, where nutrients are absorbed. Furthermore, in foregut fermenters, microorganisms from the rumen are themselves broken down by acids in the true stomach (abomasum). The resulting material, which contains the carbohydrates and proteins synthesized by the microorganisms as well as the products of fermentation, moves into the small intestine and colon. Finally, the microorganisms in the rumen can detoxify many harmful alkaloids in the plants that foregut fermenters consume. By contrast, in hindgut fermenters, food passes rapidly into the small intestine and is then mixed with microorganisms in the cecum. These animals do not digest the microorganisms present in the cecum and thus cannot exploit this potential source of nutrients. In addition, hindgut fermenters must absorb toxic plant chemicals into the bloodstream and transport them to the liver for detoxification or sequestration. Yet, although efficiency may be the trademark of foregut fermenters, hindgut fermenters are able to process material much more rapidly. For example, food moves through the gut of a horse in about 30 to 45 hours, whereas it may take a cow 70 to 100 hours to process food. Hindgut fermenters efficiently digest food high in protein, because large volumes of food can be processed rapidly. Furthermore, hindgut fermentation is effective when forage is dominated by indigestible materials such as silica and resins, because these compounds move quickly through the alimentary canal, bypassing the cecum. In sum, due to their lower efficiency, hindgut fermenters must eat large volumes of food in a short time. The foregut fermentation system is comparatively slow because food cannot pass out of the rumen until it has been ground into very fine particles. Thus ruminants do poorly on forage containing high levels of resins and tannins, because these compounds inhibit the function of microorganisms in the rumen. Furthermore, plants with high silica content break down slowly and thus impede movement of food out of the rumen.

As we've seen in this discussion of folivores, the digestive physiology of herbivores influences their ecology and distribution in several ways. Ruminants benefit most from foods that require an optimally efficient digestive system, whereas the best forage for hindgut fermenters is food that facilitates speed of digestion. Each strategy has advantages for survival in particular ecological niches. Speed of digestion may not be important to ruminant artiodactyls. When food is available in the form of tender, short herbage with high protein content, ruminants' digestive efficiency pays off. The process of rumination also permits animals to feed quickly and then move to safe cover to chew the cud at leisure.

Gliding Membranes and Pectinate Teeth

Colugos of Southeast Asia are experts at processing their leafy foods. Colugos (order Dermoptera) are represented by a single family (Cynocephalidae) containing two genera and two species (Stafford 2005). Dermopterans (meaning "skin-winged") are commonly called "flying lemurs" or colugos (fig. 3.18A), the latter name being preferable because dermopterans do not fly and are not lemurs! Colugos glide (they are "glissant"). Like the treeshrews, this order also has a confusing taxonomic history. Historically, colugos have been grouped taxonomically with the bats, "insectivores," and primates—McKenna and Bell (1997) considered dermopterans a suborder of primates. Colugos are about the size of a small house cat, weighing between 1 and 2 kg (2.2–4.4 lb), with head and body length of 340 to 400 mm (13.4–15.7 inches) and tail length of 170 to 270 mm (6.7–10.6 inches). Of the two species, the Malayan colugo (*Cynocephalus variegates*) is the larger, occurring in tropical rainforests and rubber plantations of Malaya, Thailand, Tenasserim (in Myanmar), Sumatra, Borneo, Java, and adjacent islands. The slightly smaller Philippine colugo (*C. volans*) inhabits montane forests and lowlands on the Philippine islands of Mindanao, Basilan, Samar, Leyte, and Bohol.

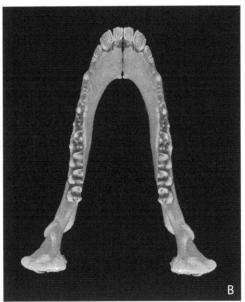

FIGURE 3.18. (A) A colugo of southeast Asia, the Sunda flying lemur (*Galeopterus variegates*). (B) Lower jaw of a colugo, showing teeth well adapted for folivory. Note the unique pectinate (comblike) incisors. These specialized teeth may be used to scrape food or strain sap, and even for grooming their lush pelage. *Sources:* (A) C. J. Phillips, ASM image library; (B) University of Michigan, Animal Diversity Web, http://animaldiversity.org, courtesy of Phillip Myers.

Colugos are primarily nocturnal, foraging on flowers, leaves, and fruits. The mean duration of foraging bouts in the Philippine colugo was found to be about 10 minutes, with 12 bouts per night being typical (Wischusen and Richmond 1998). Unlike other arboreal folivores—such as koalas (*Phascolarctos cinereus*) and three-toed sloths (*Bradypus variegates*), which consume the leaves of only one species each night—Philippine colugos are generalized foragers, feeding on young leaves from several different tree species every night. In the study region near the town of Cabarisan, Davao City, Mindanao, colugos preferred eating several species of young leaves that were plentiful and available year-round. This preference for young leaves is not surprising in the light of their higher nutritional value than older leaves. The colugos use their enlarged tongue and specialized lower incisors to pick bunches of leaves from trees. Leaves are finely masticated by "insectivore"-like teeth, accompanied by a shearing action. The stomach is specialized for ingesting large quantities of leafy vegetation and has an extended pyloric region. Colugos have an elongated intestine and large, compartmentalized cecum harboring microorganisms that break down cellulose, greatly enhancing assimilation efficiency.

Colugos den in tree cavities or hang upside down from branches during the day (with their head remaining upright, unlike bats). Colugos are helpless on the ground but adept at climbing high into trees, aided by their long, curved claws that assist in anchoring them as they rest on the trunk of a tree. Their mottled, brownish gray and white pelage camouflages them against tree trunks. Their arboreal abilities are less than acrobatic as they ascend trees in a series of slow hops; arboreal locomotor ability is compromised by their wide patagium and nonopposable thumbs. By day, colugos nest in holes or hollows of trees or hang beneath a branch, as do South American sloths.

Although their large patagium may inhibit scurrying around on tree trunks, colugos are unsurpassed in the art of gliding. Their patagium is more extensive than in any other gliding mammal. They move among trees very efficiently—their morphological adaptations for this means of locomotion are pronounced. Large eyes and stereoscopic vision enhance depth perception for these nocturnal foragers. Their large patagium extends from the neck to the digits of the forelimbs, along the sides of the body and hind limbs, enclosing the tail and tips of the fingers and toes. The development of the gliding membrane in colugos is more impressive than that in other gliding mammals such as flying squirrels (order Rodentia) and gliding possums (order Diprotodontia). Unlike the colugos, other species have a pata-gium stretched only between their limbs, with the fingers, toes, and tail left free. Despite colugos' large size, glide distances of more than 100 m (109 yards) are common, with very shallow glide angles. Thus, while gliding 100 m, a colugo loses less than 10 m in elevation. Their long-distance glides also make them vulnerable to large avian predators. This selective force was probably responsible in part for the evolution of nocturnality in colugos and flying squirrels. Like bats and other gliding species, colugos have a keeled sternum that enhances muscle attachment.

Colugos' teeth are well adapted for their folivorous diet and are unlike those of any other mammal. Like ruminants, colugos have a gap at the front of the upper jaw, with all of the upper incisors at the side of the mouth. The first upper incisor is reduced, while the second (outer) upper incisor is caniniform and has two roots—a unique feature among mammals. The teeth number 34. Another distinct and unique feature of dermopterans per-tains to the first two lower incisors, which have a peculiar "comblike," or **pectinate**, structure (fig. 3.18B). These procumbent incisors each have 5 to 20 distinct tines radiating from a single root. This comblike arrangement re-sembles the teeth of mouse lemurs, galagos, and bushbabies, as described earlier. The difference is that each lower incisor of colugos may have as many as 20 tines radiating from one root, whereas in the lemurs each tine of the comb is actually a single tooth. What is the function of this fascinat-ing arrangement? This is not yet clear. Some researchers believe that colu-gos may employ these comblike structures to scrape food or strain sap or even in grooming their lush pelage (Aimi and Inagaki 1988), whereas oth-ers believe these unique structures have no specialized function (Wischusen 1990).

Rock Badgers: Gutsy Cliff-Dwellers

One group of small mammals that have mastered the art of hindgut fermen-tation is the hyraxes. Hyraxes are in the order Hyracoidea, which comprises three genera and four species, all in the family Procaviidae. Referred to in the Bible as "rock badgers" or "coneys," hyraxes occur in central and southern Africa, Algeria, Libya, Egypt, and parts of the Middle East, in-cluding Israel, Syria, and southern Saudi Arabia. The rock hyrax (*Procavia capensis*) is the most widely distributed, both geographically and elevationally. Members of this species inhabit rocky outcrops from sea level to 4,200 m (13,780 ft) elevation in Africa and the Middle East. The yellow-spotted rock hyrax (*Heterohyrax brucei*) is found in similar rocky habitats in northeast to southern Africa (fig. 3.19A) (Barry and Shoshani 2000). The two species of arboreal tree hyraxes (*Dendrohyrax*) inhabit forested areas of southern and central Africa up to 3,600 m (11,800 ft) elevation. Rock hyraxes are about the size of a woodchuck (*Marmota monax*), ranging in mass from 1.8 to 5.4 kg (4–12 lb). They have 34 permanent teeth. Deciduous canines may be retained in rare cases, but there is usually a diastema between the incisors and cheekteeth. The upper incisors are long, pointed, and, in males, triangular in cross section with a gap between them; incisors are crescent-shaped in females. The upper incisors are ever-growing and stay sharp because the posterior sides do not have enamel. Hyraxes inhabiting warm, arid regions have shorter, less dense pelage than tree hyraxes and rock hyraxes from high-elevation, alpine areas. Hyraxes have a prominent mid-dorsal gland surrounded by either a black or light-colored patch of hair. The gland varies in size among species, being most noticeable in the western tree hyrax (*Dendrohyrax dorsalis*) and least so in the rock hyrax.

All hyraxes are herbivorous and feed on a variety of vegetation. Although they do not ruminate, hyraxes have a unique digestive system involving one large cecum and a pair of ceca on the ascending colon (fig. 3.19E). Hyraxes consume a wide variety of plant material. Grasses and herbs make up a large part of the diet of the rock hyrax; it has **hypsodont, lophodont** dentition, well adapted for grinding abrasive plant material (Kingdon 1997b). In Serengeti National Park, rock hyraxes are known to have a varied diet, feeding on *Acacia* and *Allophylus*. In addition, they may expend more than 80% of foraging time browsing on twigs and bark of woody plants, augmented by buds, leaves, flowers, and fruits of forbs, bushes, and trees (Turner and Watson 1965; Hoeck 1982).The other hyraxes have more **brachyodont** dentition because they consume less abrasive vegetation. Yellow-spotted hyraxes in Zimbabwe feed on twigs, bark, leaves, and buds of such woody species as *Combretum molle, Elephantorrhiza goetzei,* and *Kirkia acuminate* (Barry and Shoshani 2000). The southern tree hyrax (*Dendrohyrax arboreus*) in Parc National des Volcans, Rwanda, feeds primarily on mature leaves of *Hagenia abyssinica* (Milner and Harris 1999).

Because hyraxes inhabit rocky cliffs or move through trees, good trac-

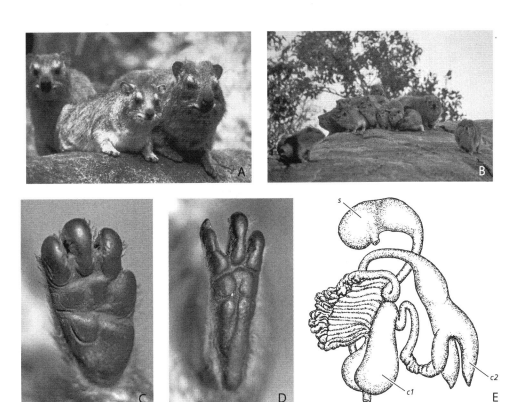

FIGURE 3.19. (A) Yellow-spotted rock hyraxes (*Heterohyrax brucei*) of Africa. (B) A group of gray hyraxes (*Heterohyrax* sp.) huddling together while "sunbathing" on rocks. (C) The left forefoot and (D) left hind foot of rock hyraxes have long nails, and the soles are equipped with glandular pads to enhance traction on rocks. (E) The alimentary canal of the rock hyrax is adaptive in breaking down its diet of fibrous plant material. The alimentary tract includes the stomach (*s*), a large cecum (*c1*) at the beginning of the large intestine, and a second, paired cecum (*c2*) at the end of the large intestine. *Sources:* (A, B) H. N. Hoeck, ASM image library; (C, D) Hendrik Hoeck; (E) Feldhamer et al. 2007.

tion for climbing and jumping has obvious adaptive value. This is achieved through specialized elastic pads on the soles of the feet (fig. 3.19C, D). These pads are kept moist by secretory glands, making the feet similar to suction cups. The toes have short, hooflike nails, except for the second digit on the hind feet, which has a claw used for grooming. Colony size varies according to species, with the rock hyrax maintaining group sizes up to about 25 and the yellow-spotted hyrax up to 34 individuals. Hyraxes are poor thermoregulators. Individuals in a colony may huddle together to help conserve heat and maintain body temperature (fig. 3.19B). Rock hyraxes bask in the sun during winter to help warm up, but during the summer they seek cool refuges in the rocks to escape lethally high temperatures and to reduce water loss (Brown and Downs 2005, 2006). Water economy is essential to cope with the extreme temperatures encountered in arid regions of the rock hyraxes' homeland. Their kidneys are efficient enough to permit them to exist on a minimal moisture intake. They have a high capacity for concentrating urea and electrolytes and excrete large amounts of undissolved calcium carbonate. Rock hyraxes establish communal "latrines" near sleeping quarters. Visible white stains result from deposits ("hyraceum") of crystallized calcium carbonate. Deposits may be centuries old. These deposits are reportedly used for medicinal purposes by several South African tribes and by Europeans. The crystals are used to treat epilepsy, hysteria, and miscellaneous injuries and as a treatment for St. Vitus's dance (a disorder of the nervous system following a group A β-hemolytic streptococcal infection).

As might be expected in colonial species, hyraxes are very vocal and make a variety of sounds, including whistles, screams, croaks, and chatter. Rock hyraxes form social hierarchies with cooperative breeding. Unlike most mammalian species, in which males have higher levels of testosterone than females, adult female hyraxes have testosterone levels equal to or greater than those of the males. Reproductive maturity in both sexes occurs at about 16 months of age. Females come into **estrus** once a year (are **monestrous**). The **gestation period** is relatively long—eight months—with litter size ranging from one to four. Neonates are **precocial**, with most births occurring during the wet season. Ron Barry (2009) has observed seasonal variation in the extent to which rock hyraxes and yellow-spotted hyraxes associate in Zimbabwe, where the two species are more closely associated than elsewhere and occupy communal nurseries in the birth season. Young males disperse from their natal area between the ages of 16 and 30 months and try to establish their own breeding territories.

CHAPTER FOUR

Carnivory

....

The **carnivorous** ("meat-eating") mammals feed primarily on animal material. Members of this group comprise the flesh- or meat-eating members of the order Carnivora and **marsupial** dasyurids (order Dasyuromorphia, family Dasyuridae). I also discuss in this chapter the **adaptations** of piscivory ("fish-eating") and sanguinivory ("blood-eating"), occurring in certain chiropterans and in the marsupial water opossum, or yapok, as a form of carnivory.

The well-known mammalian order Carnivora includes the dogs, cats, hyenas, bears, raccoons, weasels, skunks, civets, and allies. A great deal of diversity exists among the 15 families and approximately 286 species now recognized in this order. Most members of the Carnivora are too large to be labeled "small mammals," the exceptions being some of the small felids, canids, procyonids, mustelids, civets, and allies. Not all members of Carnivora eat meat. Although the cats (Felidae) and weasels (Mustelidae) are principally carnivorous, most of the canids (Canidae) are opportunistic and will eat all kinds of vertebrates, mollusks, crustaceans, insects, vegetable matter, and carrion. Most of the procyonids and ursids (bears) are notably **omnivorous**. Members of the family Ursidae are too large to be considered small mammals so are not discussed here. Scavengers, "insectivores," and vegetarians are also included in the order Carnivora.

General Characteristics
....

Carnassial Teeth

The order Carnivora is represented by a diverse array of feeding types and dental morphologies. Most carnivores are predators, typified by strong skulls, jaws, and teeth—namely, sharp **incisors** and **canines**—designed to kill and dismember prey. Carnivores can capture prey by using their large, strong, pointed canine teeth. Most carnivores also have a pair of **carnassial** teeth, a combination in which the last upper **premolar** and the first lower **molar** form a shearing surface when the mouth is closed (see fig. 3.1A). The carnassial pair is most highly developed in the cat family and least developed in the more omnivorous families of bears and raccoons. The mustelids are obligatory meat-eaters with impressive carnassial teeth. Animal material is mostly protein and is converted to energy more efficiently than plant material. Thus, as in the **insectivorous** mammals (see Chapter 2), the alimentary canal of carnivorous mammals is short, and the **cecum** (appendix) is small or absent (see Feldhamer et al. 2007: fig. 7.2).

The jaw muscles of carnivorous mammals differ from those of **herbivores** (see fig. 3.1) in the relative importance of the three major adductor muscles of the mandible: the **temporalis, masseter,** and **pterygoideus** (Elbroch 2006). Carnivores must first seize and hold their prey with the canines, which requires a large force at the front of the jaws. The very large temporalis muscles function in holding the jaws closed and aid in the vertical chewing action. The masseter is comparatively small and serves to stabilize the articulation of the jaw, and the pterygoideus helps position the carnassials. In contrast, herbivores rely heavily on a large masseter muscle to maintain a horizontal movement of molars for grinding fibrous food.

Flesh-Eating Carnivores

Felids

The cat family (Felidae) combines carnassial teeth, acute senses of smell and hearing, and impressive vision to optimize the capture and consumption of prey. Their eyes, larger than those of most carnivores, face forward, thus providing exquisite binocular vision and depth perception—both vital to locating prey. Their long, stiff, highly sensitive vibrissae (whiskers) are especially useful for foraging at night. Cats have long, sharp, usually retractile claws that serve well as effective meat hooks for capturing, slashing, and manipulating prey. They typically kill by a powerful bite to the head or neck, sometimes by suffocation. Cats use their long, sharp canines for grasping prey and their well-developed carnassials for shearing food. The dorsal surface of the tongue is covered with posterior-directed papillae that give the tongue a "sandpaper" feel and are well suited for holding prey and scraping meat from a carcass and may help to retain food in the mouth.

The family includes 14 genera and 40 species that occur worldwide except for Australia, Madagascar, and Antarctica (Wozencraft 2005; Feldhamer et al. 2007). Cats are major predators in tropical rainforests throughout the world. All cats are characterized by a shortened **rostrum**, well-developed carnassials, and large canine teeth that are highly specialized for delivering an aimed, lethal bite: the canines enter the neck between the vertebrae and separate the spinal cord. The cheekteeth are bladelike and adapted for seizing and slicing meat from prey, rather than crushing bones, as occurs in hyaenids and canids. Body mass ranges from about 1 to 2 kg (2.2–4.4 lb) in the tiny black-footed cat (*Felis nigripes*) of Africa (fig. 4.1A) to 300 kg (661 lb) in tigers of the Indian subcontinent and Russia. Large felids are noted for their ability to roar (because flexible cartilage replaces the hyoid bone at the base of the tongue), whereas smaller species purr. Most cats are at least semiarboreal. They are **digitigrade** (walking on the digits) and have strongly curved, sharp claws to hold prey. Other than in the cheetah (*Acinonyx jubatus*), claws are retractile; in the resting position the claws are held "in."

Felids prey almost exclusively on mammals and birds, although the Asian flat-headed cat (*Prionailurus planiceps*) and fishing cat (*P. viverrinus*) consume fish, frogs, and even mollusks. Most felids are **nocturnal**. They are very agile and either stalk their prey or pounce from ambush. Their **pelage** is often spotted or striped, an adaptation to cryptic hunting behavior and habitat. The cheetah is the fastest mammal in the world and relies on its ability to outrun prey over short distances. Felid species generally are solitary or form pairs; however, lions (*Panthera leo*) associate in prides that include up to 18 related females, their offspring, and several unrelated males.

There are 12 species of cats that qualify as "small mammals." The black-footed cat (*F. nigripes*), mentioned earlier, and the rusty-spotted cat (*Prionailurus rubiginosus*) are the smallest cats. About the same mass as the black-footed cat is the rusty-spotted cat, which also weighs between 1 and 2 kg. This small cat lives in scrub forests of southern India and Sri Lanka, where it preys on insects, birds, and small mammals. Other notable small cats of the Old World include the sand cat (*Felis margarita*), a nocturnal desert dweller of North Africa and Southwest Asia; the wildcat (*F. silvestris*) of Europe; and the bay cat (*Pardolfelis badia*) of Borneo.

FIGURE 4.1.   (A) Arguably the smallest of all cats, the tiny black-footed cat (*Felis nigripes*) of Africa weighs between 1 and 2 kg (2.2–4.4 lb). (B) A close second to the black-footed cat in terms of small size is the sand cat (*Felis margarita*), found in the Sahara Desert. South America is home to 10 species of "small cats." Two representatives are (C) the oncilla (*Leopardus tigrinus*) and (D) the margay (*Leopardus wiedii*). *Sources:* (A) J. Visser, ASM image library; (B) Susan M. G. Hoffman; (C, D) Rexford D. Lord.

The tiny black-footed cat inhabits arid, sandy regions of South Africa, Botswana, and Namibia. German scientist Alex Sliwa (1994) carefully observed the foraging behavior of the black-footed cat. Of more than 2,000 captures of prey, the menu was found to be headed by large-eared mice (*Malacothrix typica*), along with other small rodents, shrews, and 21 species of birds (Hunter 2005: 77). These small cats were reported to consume prey as large as cape hares (*Lepus capensis*) weighing up to 3.5 kg (7.7 lb) and were observed scavenging on dead springboks (*Antidorcas marsupialis*). Sliwa noted that black-footed cats **cached** partially eaten prey in shallow aardvark digs, concealing the cache with a covering of soil and grass. Cats did not return to a specific cache immediately, and they continued hunting even when many kills were stockpiled. These small cats are faced with winter night-time temperatures below −8°C (17.6°F) and have high metabolic requirements. Their immediate energetic needs are fulfilled by continuous hunting, while established caches act as a nutritional backup if needed.

The northern equivalent of the black-footed cat is the sand cat (*F. margarita*) of the Sahara Desert (fig. 4.1B). These are very small, sandy-yellow cats weighing about 1.5 to 3.4 kg (3.3–7.5 lb). They have grayish freckling on the flanks, shoulders, and forehead and dense, fine fur. For their size, sand cats have exceptionally long canines, and they are perhaps the most **fossorial** member of the cat family. The principal prey of these tiny cats are gerbils and lizards. Little direct evidence is available on specific techniques of capture of prey by sand cats; however, local people have observed their rapid burrowing technique when excavating prey in the desert environment. The Touareg people of Niger offer praise for the digging skills of the sand cat by referring to it as "the cat that digs holes" (Hunter 2005: 62). Local people also regard the sand cat as a snake specialist renowned for killing horned vipers (*Cerastes cerastes*) and sand vipers (*C. vipera*) of the Sahara Desert. Both vipers are highly venomous. A sand cat first teases a snake with a succession of bites to the head until the prey becomes vulnerable. Next, the cat pins down the exhausted snake with its well-furred paws while inflicting many bites to the skull or neck. It then devours the entire viper, consuming venom glands and fangs without harm (Hunter 2005: 77). Well adapted for life in the desert, sand cats gain most of the moisture they require from prey and they conserve water by entering **torpor** during the day, sequestered in their deep burrow.

A plethora of small cats reside in the neotropical region. The formation of the land bridge between North and South America set the stage for a fascinating biogeographic event. This land bridge, forming some 3 to 4 million years ago, greatly favored immigration and colonization of South America by small northern felids. Today, 10 species of "small cats" reside in South America, of which 6 species weigh less than 5 kg (11 lb). The small cats of South America include the oncilla (*Leopardus tigrinus*) and margay (*L. wiedii*) (fig. 4.1C, D), Andean mountain cat (*L. jacobita*), kodkod (*L. guigna*), colocolo (*L. colocolo*), Geoffroy's cat (*L. geoffroyi*), and pampas cat (*L. pajeros*). The remaining three species—cougar (*Puma concolor*), ocelot (*Leopardus pardalis*), and jaguarondi (*Puma yagouaroundi*)—are too large to qualify as small mammals. The jaguar (*Panthera onca*)—one of the "big cats"—is the largest cat of the New World, weighing up to 113 kg (249 lb). Margays, slightly larger than a house cat (body mass 3–6 kg, or 6.6–13.2 lb), inhabit mature and secondary growth evergreen and deciduous forests of Central and South America, from Mexico south to Uruguay and Argentina (De Oliveira 1998). Margays are well adapted to **arboreal** life; their paws are wide and flexible, with supple digits and large claws. These are the only New World felid with ankle joints that have the ability to rotate 180° around their longitudinal axis, allowing the cats to descend a tree head down, like a squirrel. Their diet is diverse, including arboreal mammals, birds, amphibians and reptiles, insects, and fruits. Slightly smaller yet similar in appearance to the margay is the oncilla (*L. tigrinus*). Oncillas, weighing 1.5 to 3 kg (3.3–6.6 lb), are found in brushy areas and evergreen and deciduous forests of Central and South America. Oncillas prey on small mammals and birds as well as small primates. There is a great deal of disagreement concerning the classification of the genera of neotropical cats (Emmons 1997).

Mustelids

Members of the family Mustelidae, equipped with prominent sharp canines and well-developed carnassials, have a reputation as some of the fiercest of carnivores. With short jaws and sharp teeth, mustelids are second only to the felids as **specialists** in flesh-eating abilities. The most important teeth are the four carnassials (Elbroch 2006). They are strategically located at the rear of the jaw so as to take the utmost advantage of the leverage of the jawbone and the strength of the huge temporalis muscles. This arrangement explains why weasels, like canids, chew bones at the corner of their mouth (C. M. King 1990: 19). These small carnivores have an extremely powerful bite for their size, and some kill prey much larger than themselves. The Mustelidae family—the largest family of order Carnivora—is large and diverse, with 22 genera and 59 species that include weasels, badgers, otters, and the wolverine (*Gulo gulo*). Mustelids are found on all continents except Antarctica and Australia. Members of this family inhabit both **terrestrial** and arboreal habitats as well as freshwater and saltwater habitats. They range in size from the least weasel (*Mustela nivalis*), weighing 30 to 70 g (1–2.4 oz), the world's smallest carnivore (fig. 4.2A), to the wolverine (*G. gulo*), at 55 kg (121 lb). Male mustelids generally are about 25% larger than females. Mustelids have long bodies with relatively short legs. Their slender bodies are advantageous when searching for food, as they can track prey seeking refuge belowground or underneath snow or high in trees. This long, skinny

FIGURE 4.2. Two common members of the family Mustelidae living in North America are (A) the smallest of all carnivores, the least weasel (*Mustela nivalis*), and (B) the semiaquatic American mink (*Neovison vison*). The family Canidae includes (C) the fennec (*Vulpes zerda*) of northern Africa, the smallest of all canids, weighing 0.8 to 1.9 kg (1.7–4.2 lb); and (D) similar in size, Blandford's fox (*Vulpes cana*), which inhabits the mountainous regions of Israel and the Arabian Peninsula eastward to Afghanistan and Pakistan. *Sources:* (A, B) James F. Parnell; (C) Chuck Dresner, courtesy of Saint Louis Zoo; (D) Chris and Mathilde Stuart, African-Arabian Wildlife Research Centre.

body may be adaptive for stalking prey, but much energy is lost as heat due to its large surface-to-volume ratio. To stay ahead of rapid heat loss, mustelids have high **metabolic rates** and thus very high food requirements. Mustelids have acute senses of smell and hearing.

Mustelids are active, fierce hunters, many with specialized methods of killing prey (C. M. King 1990; Powell 1993). Although most are small, size does not seem to hamper their foraging ability. For example, least weasels (*M. nivalis*) have a circumboreal (northern **circumpolar**) range throughout the Holarctic. Least weasels are found in Pennsylvania, for example, but are not very common; their optimal habitat is brushy areas, open woodlands, and old field and pasturelands that support high numbers of prey species of small mammals (principally meadow voles and white-footed mice). Because of their nocturnal habits, small size, and secretive nature, least weasels are rarely observed; however, from my experience, house cats have minimal difficulty finding them and occasionally dropping them at the doorstep. These small mustelids (their body is about the thickness of a man's thumb) are ferocious hunters, consuming up to one-half their body mass in food each day, and thus can significantly affect a population of small mammals. For its body mass, the least weasel has the most powerful jaws of any predator in North America. Christiansen and Wroe (2007: table 1) calculated the bite force of some 190 species of extant and fossil mammals. Compared with other extant species, the average bite force for *M. nivalis* was 164, second only to that of the Tasmanian devil (*Sarcophilus harrisii*), with a bite force of 181 (Wroe, McHenry, Thomason 2005: table 1).

Voles make extensive tunnels through the grass and under the snow, and weasels are the supreme experts at tunnel hunting. In southern Ontario, Simms (1979: 515) determined that female ermines were "optimally sized to forage in the subnivean environment" and could gain access to more than 90% of **subnivean** tunnels of meadow voles (*Microtus pennsylvanicus*). The ermine's killing technique is typified by seizing the victim by the head and neck and wrapping its sinuous body around the prey, which is most likely larger than the predator. It then kills its prey by inflicting a rapid bite to the base of the skull or by severing the jugular vein with its sharp teeth. It first consumes the brain, then moves on to the heart, lungs, and, ultimately, the entire body, including most bones and fur. Least weasels (or for that matter any of the weasels) do not suck blood but commonly lap fresh blood seeping from the rear of the victim's skull. This habit of lapping or licking blood from inflicted wounds may have given rise to the false idea that weasels "suck the blood" of their prey. A prey animal is commonly devoured on the spot but occasionally may be cached near the den or taken to the nest to feed young.

There are many eyewitness accounts of weasels attacking prey larger than themselves. For example, Bachman reported of the ermine that "this little Weasel is fierce and bloodthirsty, possessing an intuitive propensity to destroy every animal and bird within its reach, some of which, such as the American rabbit, the ruffed grouse, and domestic fowl, are ten times its own size. It is a notorious and hated depredator of the poultry house, and we have known forty well grown fowls to have been killed in one night by a single Ermine" (Audubon and Bachman 1849, 2: 58).

Weasels are said to be "the most ferocious and aggressive of all our predatory mammals" (Oehler 1944: 198). Charles Oehler documented an interaction with an aggressive long-tailed weasel in the autumn of 1940. While collecting beetles near his home in Cincinnati, Ohio, Oehler and a companion noticed an animal dashing into a hollow log. Curiosity drove them to pound on the log in the hope of capturing the animal (suspected to be a chipmunk). Now comes the best part:

> We had scarcely begun, when the infuriated little beast dashed madly from its retreat, teeth bared for action. Before a capture could be executed, it was chewing on my companion's hand. I rushed to his assistance and, grabbing the animal by the neck, was forced to pry its teeth from his bleeding hand. I had no sooner freed my companion, than I realized I had a double hand full of trouble. My would-be captive seemed to have more control of the situation than I, chewing and clawing in a most determined manner. The only reason I did not free the animal was because it would not cooperate. By the time the weasel was subdued (30 minutes later, with the help of a small club and both of my feet) my hands were a sorry sight indeed. I have never witnessed, or been the victim of, a more remarkable exhibit of ferocious aggressiveness and stamina. (Oehler 1944)

This is not an uncommon occurrence. In conversations with my colleagues, I have learned that others have experienced such "difficulties" with small mustelids—they are clearly a "no-nonsense" group of carnivores.

Caching behavior is well documented for weasels. For example, when investigating below the floorboards of a barn, Bachman observed "an immense number of Rats dragged together forming a compact heap" (Audubon and Bachman 1849, 2: 60). Piles of 100 or more rats and mice have been reported (Audubon and Bachman 1849, 2: 60; Seton 1929: 595). Gerry Svendsen observed a female long-tailed weasel hunting in an alpine meadow in Colorado: the weasel "found two nests of one- to two-week-old golden-mantled ground squirrels (*Spermophilus lateralis*). She killed all nine young in the burrow and carried them to an abandoned pocket gopher burrow where they were cached. The weasel visited the cache several times in the next few hours and added new items to the same cache during the next 2 weeks. This was a **lactating** female weasel that had a burrow containing young located about 175 meters from the cache site" (Svendsen 1982: 623). Caching behavior is adaptive when a locally abundant food source is discovered. Prey can be quickly killed and then consumed later, thus conserving energy used in searching for food.

One of the most successful mustelids is the American mink (*Neovison vison*) (fig. 4.2B). The Reverend John Bachman was not enamored with the mink, as he notes: "Next to the ermine, the Mink is the most active and destructive little depredator that prowls around the farm-yard, or the farmer's duck-pond; where the presence of one or two of these animals will soon be made known by the sudden disappearance of sundry young ducks and chickens" (Audubon and Bachman 1849, 1: 254). The American mink is found throughout Canada and most of the United States, except in dry parts of California, Nevada, Utah, New Mexico, Arizona, and western Texas (Larivière 1999). In North America, the adaptability of the American mink is clearly reflected in the great variety of habitats in which it thrives. American mink were first taken to Europe in the 1920s to set up fur farms, and escapees from the farms established populations throughout much of Europe. Introduced for fur farming in the 1930s, mink are currently colonizing parts of South America. American mink have a varied carnivorous diet governed by seasonal availability of prey. During summer in Pennsylvania, for example, crayfish top the menu, followed by muskrats, frogs, fish, snakes, small mammals, and waterfowl, especially flightless young and molting adults. During winter, muskrats are a staple. Because *N. vison* spends more time foraging away from water at this time, it also commonly feasts on small mammals such as voles, mice, shrews, cottontail rabbits, and, occasionally, squirrels. A skillful and dedicated hunter with poor senses of sight and hearing, the mink relies on its keen sense of smell to locate prey. Its killing tactics are similar to those of other weasels: it inflicts a series of deadly bites to the neck and base of the skull. If the mink kills more prey than it can consume on the spot, it carries the victim by the neck to a den and stores the carcass for winter. Such caches may be quite large; one biologist reported a cache containing more than a dozen muskrats, two mallard ducks, and an American coot (Yeager 1943).

Canids

The family Canidae includes wolves, coyotes, foxes, dingoes, dholes, jackals, and dogs. The family consists of 13 genera and 35 species that occur naturally throughout the world except in Antarctica. The dingo (*Canis lupus dingo*) was introduced to Australia, New Guinea, and parts of Asia 3,500 to 4,000 years ago (Corbett 1995). Habitats of canids range from hot, dry deserts to tropical rainforests to arctic ice. Canids are opportunistic hunters that rely on high intelligence, social organization, and superb behavioral adaptability. Small canids hunt singly or in pairs, whereas larger canids, such as gray wolves (*Canis lupus*), hunt in packs of up to 30 members seeking prey that are far larger than themselves. Canids generally consume animal prey throughout the year; however, plant material may be taken seasonally by some species. Jackals often eat carrion, and the bat-eared fox (*Otocyon megalotis*) consumes large quantities of insects.

Wild canids range in size from the tiny fennec, the size of a Chihuahua dog, to the massive gray wolf of North America and Asia, which reaches a mass of 80 kg (176 lb). Structurally, canids are doglike in appearance, with elongated muzzles, long legs, and bushy tails. The long nose of canids houses complex turbinal bones responsible for their keen sense of smell. Canids are highly intelligent. They are generally digitigrade and have non-retractile claws. Canid skulls characteristically have an elongated rostrum, with well-developed canines and carnassial teeth endowed with effective cutting or shearing surfaces (Elbroch 2006). Unlike felids and mustelids, canids do not have a "killing bite" and they kill mammalian prey by shaking it and breaking the back or disabling the victim with bites to the legs, nose, and venter. Their molar teeth have crushing surfaces, an adaptation that allows them to eat almost anything. The smallest of canids (<4.5 kg, or 10 lb) include bat-eared foxes (*Otocyon megalotis*) inhabiting eastern and southern Africa; fennec and Rüppell's foxes (*Vulpes zerda* and *V. rueppellii*) of North Africa, Sudan, and Somalia; Blandford's fox (*V. cana*) of western and southern Asia; swift and kit foxes (*V. velox* and *V. macrotis*) of deserts of western and central North America; and the arctic fox (*V. lagopus*) of the circumpolar arctic tundra.

The fennec is the smallest of all canids, weighing 0.8 to 1.9 kg (1.7–4.2 lb) (fig. 4.2C). It is distinguished by its huge ears (**pinnae**), which are the largest in the dog family relative to body mass and act as "radiators" to dissipate heat in the harsh desert environment of northern Africa. These diminutive canids sport a cream-colored coat that is long, woolly, and soft, coupled with a black-tipped tail. They have short legs and a small, pointed muzzle. Fennecs reside in sand-dune deserts and steppes typified by light, sandy soils. These nocturnal canids feed primarily on grasshoppers and locusts, lizards, small mammals, birds, and roots. They hunt alone, locating prey primarily by sound; excavation is done by digging with all four feet. Detection of burrowing insects and small fossorial mammals is enhanced by the fennec's large ears and inflated **auditory bullae.** Fennecs use their slender canines to kill prey by a rapid bite to the neck; the heads of mammals are consumed first. Entire birds, including the feathers, are eaten. Extra food is cached in a manner similar to that used by larger foxes such as red foxes (*V. vulpes*), in which a hole is excavated, food deposited, and sand pushed over the food with the nose. Inhabiting the same geographic areas as the fennec is Rüppell's fox (*V. rueppelli*). Rüppell's fox is slightly larger than the fennec (weighing about 1.5 kg, or 3.3 lb) and inhabits dry sand and stone deserts of northern Africa, south to Sudan and Somalia (Larivière and Seddon 2001). These foxes are omnivorous and opportunistic predators, taking advantage of the available prey. They are primarily insectivorous, but their diet also includes rodents, birds, lizards and fruits. They are reported to climb date palms and will forage on fibrous fruits of dom palms (Osborn and Helmy 1980).

One of the rarest predatory mammals in southwestern Asia is Blandford's fox (*V. cana*) (fig. 4.2D). Similar in size to a fennec, Blanford's fox inhabits mountainous regions of Israel and the Arabian Peninsula, eastward to Afghanistan and Pakistan, with isolated populations in eastern Egypt. This handsome little canid (body mass 0.8–1.5 kg, or 1.8–3.3 lb) sports a brownish gray pelage and is equipped with a magnificent tail that almost equals its body length. Blandford's foxes in Israel are primarily insectivorous and **frugivorous.** Invertebrates such as beetles, grasshoppers, ants, and termites are a staple in their diet. Plant food consists mainly of two species of caperbrush (*Capparis cartilaginea* and *C. spinosa*) along with other locally abundant plants and fruits (Geffen et al. 1992; Geffen 1994). Although most canids tend to be **cursorial** (running locomotion), Blandford's foxes, like the common gray fox (*Urocyon cinereoargenteus*), are highly arboreal, using their long, bushy tail as a counterbalance during foraging. Naked footpads and sharp, curved claws enhance traction on vertical surfaces. Blandford's foxes are strictly nocturnal and almost always forage alone (Geffen et al. 1992).

Kit and swift foxes (*V. macrotis* and *V. velox*) are the smallest wild canids in North America, weighing between 1.6 and 2.7 kg (3.5–5.9 lb). Kit foxes closely resemble swift foxes in appearance and habits. Kit foxes occupy semidesert shrubland and margins of pinyon-juniper woodlands of the southwestern United States and northwestern Mexico (McGrew 1979). Salt-brush, shadscale, sagebrush, and greasewood are common woody plants in the habitats of kit foxes. Swift foxes inhabit short, medium, and mixed grass prairies of the west-central United States, with populations extending into southern Canada (Cypher 2003). They occur side by side with black-tailed prairie dogs (*Cynomys ludovicianus*) and white-tailed jackrabbits (*Lepus townsendii*).

Kit foxes and swift foxes primarily consume animal prey. For example, in cold desert regions of Utah, kit foxes consume mainly black-tailed jackrabbits (*Lepus californicus*) and desert cottontails (*Sylvilagus audubonii*). In Mexico, prairie dogs (*Cynomys ludovicianus*) are the main prey item. Food habits vary greatly with availability of prey. Jackrabbits are also a staple in the diet of swift foxes, along with ground squirrels, prairie dogs, and a variety of ground-nesting birds. Other items commonly consumed include thirteen-lined ground squirrels (*Spermophilus tridecemlineatus*), deer mice (*Peromyscus*), voles (*Microtus*), pocket mice (*Perognathus*), and grasshopper mice (*Onychomys leucogaster*), to mention just a few. Kit foxes and swift foxes do not consume very much plant material. Like other fox species, both kit and swift foxes readily cache food.

Piscivores

Claws, Cheek Pouches, and a Sixth Finger

Some predators are **piscivorous** ("fish-eating"). At least four species of bats are known to hunt for prey by aerial hawking and by capturing prey from the surface of the water by using their feet or **uropatagium** (trawling). Species include the greater bulldog bats (*Noctilio leporinus*), fish-eating myotis (*Myotis vivesi*), large-footed myotis (*M. adversus*), and Daubenton's bat (*M. daubentonii*). Piscivory most likely evolved in insectivorous bats that utilized trawling. Perhaps the best example of piscivorous bats is found in the New World family of bulldog bats, Noctilionidae (fig. 4.3A). This family inhabits tropical lowlands from northern Mexico to northern Argentina and on several Caribbean islands. The greater bulldog bat (*N. leporinus*) has a head and body length up to 13 cm (5 inches). Males may weigh 70 g (2.5 oz) and have a 50 cm (19.5 inches) wingspan. Pelage is extremely short and ranges in color from pale orange to brownish and even grayish brown.

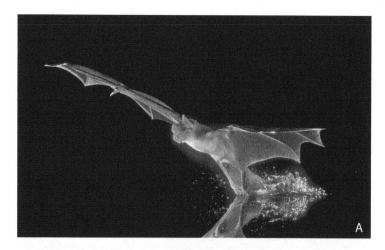

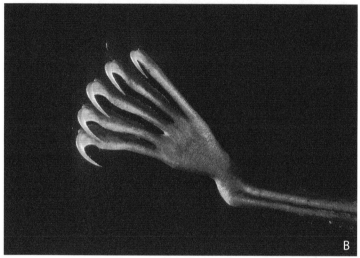

FIGURE 4.3. (A) The greater bulldog bat (*Noctilio leporinus*) of the family Noctilion-idae captures prey from the surface of the water by using its feet or uropatagium. (B) Bulldog bats are characterized by unusually long, robust hind limbs and large feet with sharp, recurved, laterally flattened claws, which they use to scoop up small fish, crustaceans, and aquatic insects. *Source:* Merlin D. Tuttle, Bat Conservation International.

The lesser bulldog bat (*N. albiventris*) inhabits the same geographic region but is smaller than *N. leporinus*—about 8 to 9 cm (3.1–3.5 inches) in head and body length, with a minimum body mass of about 40 g (1.4 oz) (Hood and Knox Jones 1984). Its limbs, particularly the femur, are short and not as robust as those of *N. leporinus*. The lesser bulldog bat feeds primarily on insects, which are captured using **echolocation**.

Greater bulldog bats are remarkable for their structural and behavioral modifications for capturing and consuming fish. The vernacular name "bulldog bat" stems from their large, swollen jowl-like upper lips that impart a bulldoglike appearance. *N. leporinus* is characterized by unusually long, robust hind limbs and large feet equipped with sharp, recurved, and laterally flattened claws (fig. 4.3B), which are used to scoop up small fish, crustaceans, and aquatic insects (Brooke 1994; Bordignon 2006). These bats hunt at dusk and during the night, typically in small groups. Vision and olfaction are not necessary to locate prey; bulldog bats use echolocation to detect ripples caused by fish swimming near the surface of the water. They skim low, dragging their elongated feet through the water, with limbs and hooklike claws rotated forward, thus acting as a gaff (fishing spear). Their long, sturdy **calcar** can be folded forward to lift the tail membrane clear of the water as they swoop to lift out the fish. Once gaffed, the fish is quickly transferred to the mouth, where it is partially masticated by the bat's formidable dental arcade and stored in well-developed cheek pouches (Schnitzler et al. 1994). Fish are either eaten in flight or carried to the bat's roost, where they are eaten while the bat is resting. Fish up to 8 cm (3.1 inches) long may be captured, and 30 to 40 fish may be taken per night. *Noctilio*'s stomach has a unique modification of the cardiac sphincter at the junction of the esophagus that permits storage of large food items. The bat usually forages over small pools, slow-moving rivers, or sheltered lagoons. Bulldogs bats have also been observed foraging over the sea at the edge of the surf, catching fish disturbed by pelicans. Although fish are the staple in the diet of *N. leporinus*, these bats also consume large amounts of insect prey. The hind limbs of the other fish-eating bats are not as well developed as in the greater bulldog bats. *N. leporinus* feeds primarily on insects using echolocation. Colonies are reported to house 75 to several hundred individuals. Hollow trees and sea caves are popular roosting sites.

Another piscivorous bat found in the New World is the fish-eating myotis (*M. vivesi*) in the family Vespertilionidae. *M. vivesi* is the largest member of the genus *Myotis* in the New World, averaging about 25 g (0.88 oz) in body mass and with a wingspan of about 40 cm (15.7 inches). This species is found primarily on islands in the Gulf of California and coastal Mexico (Blood and Clark 1998). With its long legs and huge feet equipped with long, sharp claws adaptive for gaffing fish, *M. vivesi* represents an excellent example of evolutionary **convergence**. Greater bulldog bats and fish-eating myotis have wings with a high aspect ratio (wing tip is relatively elongated and rounded), enhancing flight efficiency over water, and low wing loading, an adaptation for slow flight and carrying large prey (Altringham 1996). Fish-eating myotis employ a different prey-capture technique than *Noctilio*. High-speed photographic techniques have demonstrated that, unlike *N. leporinus*, which captures prey with its claws while keeping its uropatagium above the water surface, *M. vivesi* lowers the entire trailing edge of its uropatagium into the water to capture prey (Altenbach 1989). Fish-eating myotis are reported to consume small fish and marine crustaceans, and they roost in crevices and under stones along coasts of the Sea of Cortez. Their kidneys are well adapted for dealing with the high-saline, rich marine foods. The renal cortex is thin and the medulla thick, with a long loop of Henle extending into the renal papilla. This microanatomy of the kidney closely resembles that of Merriam's kangaroo rat (*Dipodomys merriami*). Such an arrangement optimizes the ability of *M. vivesi* to concentrate salt in its urine and use seawater as a water source.

Myotis has the broadest geographic distribution of any chiropteran genus; however, only one species is known to occur in Australia—the large-footed myotis (*M. adversus*). The large-footed myotis inhabits a coastal band in northwestern Australia that ranges across the northern tip of the continent and south to western Victoria. Except along major rivers, populations are found along the coast and rarely more than 100 km (62 miles) inland. Averaging about 10 g (0.35 oz) in body mass, these bats sport a grayish brown pelage. As the name implies, they have exceptionally large feet, 10 to 14 mm (0.39–0.55 inches) in length, and their wingspan is about 28 cm (11 inches). *M. adversus* forages over streams, ponds, and large lakes. As the bats swoop over the water, they use sharp, recurved claws on their large feet to rake the surface for aquatic insects and fish, which make up the bulk of their diet (Dwyer 1970).

Another piscivorous species of *Myotis* is more common, occurring throughout the northern Palaearctic region. Daubenton's bat (*M. daubentonii*) is slightly smaller than the large-footed myotis, weighing about 8 g (0.3 oz); it has a wingspan of about 25 cm (9.7 inches). Its feet are distinguished by their large size (more than one-half the length of the tibia) and very long calcar taking up two-thirds of the margin of the uropatagium. These bats are normally seen foraging over lakes, ponds, and streams. They consume primarily insects but are known to feed on crustaceans and small fish. In captivity, *M. daubentonii* is reported to consume 7 to 10 fish per day (Bogdanowicz 1994).

Jones and Rayner (1988, 1991) used multiple-flash stereo-photogrammetry and recording of echolocation calls to examine and compare the foraging tactics of the free-ranging large-footed myotis in Queensland, Australia, and Daubenton's bat in the United Kingdom. The style of trawling of *M. adversus* and *M. daubentonii* was similar to that described for *M. vivesi*. While trawling for prey, they break the surface of the water with the trailing edge of their uropatagium; this differs from the technique used by bulldog bats (*Noctilio*), which capture prey with their hind feet using the uropatagium to reduce the chance of dropping prey following capture. Although their wing shapes are different, *M. adversus* and *M. daubentonii* were found to forage in very similar ways. The slightly heavier body mass and larger wings of *M. adversus* produce a lower wing loading; nevertheless, it flies faster than *M. daubentonii* during the searching phase of foraging. The smaller *M. daubentonii* is able to turn in tighter circles than *M. adversus*.

Among mammals, bats do indeed monopolize the piscivorous niche, but there is one species of marsupial for which fish is a dietary staple—namely, the water opossum, or yapok, of the southern hemisphere of the New World. Yapoks are represented by a single species, *Chironectes minimus*, found from southern Mexico and Belize to northeastern Argentina. Yapoks typically inhabit clear, fast-flowing, freshwater streams with rocky substrates (Galliez et al. 2009). They range from 604 to 790 g (1.3–1.74 lb) in body mass (about the same as an eastern gray squirrel, *Sciurus carolinensis*) and sport a short, fine, dense coat, marbled gray and black on the dorsum, with a chin, chest, and belly of creamy white (Nowak 1999). This striking color pattern is unique among marsupials and may serve as camouflage for this semiaquatic small mammal. The dorsal bands are thought to blend with the stream ripples, giving a disruptive appearance in the eyes of potential aerial predators (Brosset 1989). Convergent with river otters (Mustelidae), yapoks are characterized by a streamlined body, large webbed hind feet, dense non-wettable fur, and supernumerary facial bristles (Marshall 1978). Yapoks are the only opossums with webbed feet, and they are the sole neotropical marsupial with an accessory opposable sixth "finger" on the forefeet, which is derived from an enlarged wrist bone. Another adaptation for aquatic life is the waterproof, rear-opening marsupium, which permits the female to swim with young in her pouch. The waterproof nature of the pouch is facilitated by a well-developed sphincter muscle that creates a watertight compartment. Young yapoks can tolerate low oxygen level for many minutes. In males, the pouch cannot be fully closed, but the scrotum is pulled up into the pouch during swimming, thus ameliorating the effects of low water temperatures on the testes (Marshall 1978).

Yapoks are excellent swimmers and divers, feeding on aquatic animals such as crayfish, shrimp, and fish. Unlike other aquatic predators such as American water shrews, which detect prey underwater using their sense of smell, yapoks locate aquatic prey by contact with their forefeet. Long fingers and sandpaper-like palms aid in probing for prey and grasping slippery fish. Interestingly, their forefeet are not used for propulsion, but rather the yapok holds its feet in front of its body while swimming.

Sanguinivores

Knifelike Teeth and Heat-Sensitive Nasal Pits

As we've seen in this and earlier chapters, bats seem to have covered all the bases when it comes to exploiting food niches. However, there is still another option when seeking nutrition: the consumption of blood. Three species of vampire bats (family Phyllostomidae) inhabit arid and humid regions of the New World, from Mexico south to northern Argentina. They typically reside in caves but also occupy hollow trees, old wells, mine shafts, and abandoned buildings. The vampire bat (*Desmodus rotundus*) preys exclusively on mammals (Greenhall, Joermann, and Schmidt 1983), but the white-winged vampire bat (*Diaemus youngi*) (Greenhall and Schutt 1996) and the hairy-legged vampire bat (*Diphylla ecaudata*) prefer avian prey (Greenhall and Schmidt 1988; Feldhamer et al. 2007).

The vampire bats, the only **sanguinivorous** ("blood-eating") mammals, provide an excellent illustration of how the morphology of teeth, alimentary canal, and limbs is strongly correlated with food habits (fig. 4.4A). *D. rotundus* is a medium-sized bat with a head and body length of 70 to 90 mm (2.7–3.5 inches) and weighing between 25 and 40 g (0.87–1.4 oz). It preys primarily on medium-sized and large terrestrial mammals. When found near human settlements, it ingests blood from cattle, horses, mules, goats, pigs, sheep, and humans. Today, vampire bats occur in close proximity to livestock farming and thus show a high degree of preference for cattle. At-

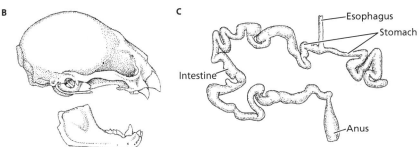

FIGURE 4.4. (A) The vampire bat (*Desmodus rotundus*) occurs only in the New World, ranging from northern Mexico to southern South America. (B) The skull of the vampire bat, showing the bladelike upper incisors and canines. The sharp points of the upper incisors fit into distinct pits in the lower jaw behind the incisors. (C) The alimentary canal of the vampire bat. The stomach serves to store large amounts of blood and absorb water rather than to digest protein, as in most mammals. *Sources:* (A) Rexford D. Lord; (B, C) Feldhamer et al. 2007.

tacks on humans do occur, often following the removal of livestock from an area. Bites to humans are not painful; however, they can be dangerous due to the possibility of transmitting paralytic rabies. Morphologically, *D. rotundus* is well adapted for a diet of blood in the following ways:

—The rostrum is reduced, supporting middle upper incisors that are triangular, forward pointing (**procumbent**), knifelike, and larger than canines. The razor-sharp points of the incisors fit into pits in the lower jaw (fig. 4.4B).

—Cheekteeth are tiny.

—The tongue has a pair of grooves at each border that function like drinking straws.

—The stomach is long and tubular, highly distensible, and well vascularized to enhance the storage of blood and absorption of water (fig. 4.4C).

—The small intestine is thin-walled and twice as long as the stomach.

—The kidneys have a unique excretory ability linked with feeding and roosting behavior.

The strong and well-developed humerus supports a thumb that is unusually long and equipped with three pads that function like a sole. The forelimbs are unique and greatly aid in terrestrial locomotion. Vampire bats are agile, quadrupedal runners and will readily run and hop on all fours, using the thickened thumb as a sole. *Desmodus* detects vascular areas of the victim by use of specialized heat-sensitive pits surrounding its nose. A bat silently approaches an animal, typically landing on the ground near its leg, and climbs or jumps to a feeding site, usually the legs, shoulders, or neck. The upper incisors and canines are used to remove a small piece of skin from the prey or make an incision several millimeters deep. Movement of blood is facilitated by several anticoagulants in the bat's saliva. The flow of blood is maintained by peristaltic waves of the tongue as the bat rapidly licks and continuously abrades the wound. Up to 13 vampire bats have been observed feeding on the neck of a cow at the same time, with a feeding time of 9 to 40 minutes. Within a three-hour period, seven bats may feed from the same wound, one after another.

The long and highly vascularized stomach of *Desmodus* does not function in protein digestion as in most mammals (fig. 4.4C). Rather, it is important in the storage of blood and the absorption of water to concentrate the blood. Average consumption of blood in the wild is about 20 mL/day. Bats may consume up to 50% of their body mass in blood per night, which might impose serious constraints on their ability to fly. Vampire bats cope with this potential problem by employing a unique "two-phase" renal function: in the first phase, which occurs at the feeding site, water is excreted; in the second phase, which takes place at the roost, urine is concentrated. About an hour after feeding, bats rapidly lose much of the water taken in with the blood meal—about 25% of the ingested blood is excreted as urine.

This weight loss is essential to enable the bats to fly back to the roost. At the roost, digestion of the partially dehydrated blood continues. Bats concentrate wastes and thus excrete a highly concentrated urine. The kidney of vampire bats may surpass that of many desert mammals in its ability to concentrate urine and thus conserve water.

Vampire bats also practice blood sharing at the roosting site by regurgitating blood into the mouth of another bat. This food sharing is a fascinating social behavior that is a rare example of reciprocal altruism. Vampire bats will starve to death if they do not eat for about three days. In roosting colonies of females (males roost individually), a bat that has recently eaten will regurgitate blood to a roost-mate that is close to starvation. Food sharing even occurs between unrelated individuals, but only between bats that have a close roosting association—that is, sharing occurs with those who can return the favor when necessary.

CHAPTER FIVE

Omnivory

....

Most mammals are **omnivorous** ("everything-eating"). Each order of mammals contains omnivorous species; however, omnivory is best illustrated in opossums, in primates such as humans and many monkeys, and in pigs, bears, and raccoons. Unlike the specialized **carnassial** teeth of canids, felids, and mustelids, the dentition of mammalian omnivores is versatile, adapted to process a variety of foods. Omnivorous mammals retain piercing and ripping **cusps** in the anterior teeth but typically have flat, broad cheek-teeth with low cusps (**bunodont** teeth) adapted for crushing food. The stomach of many omnivores, such as pigs (*Sus scrofa*), is comparatively simple. The small intestine is elongated, and the colon is large with many folds and bands of longitudinal muscle. The **cecum** of most omnivores is poorly developed, related to the small amount of fibrous plant material in the diet.

The success of raccoons (*Procyon lotor*) in North America is a good example of a mammal that combines omnivory with opportunism to enhance survival. Although raccoons show distinct food preferences, availability largely dictates food selection. During spring and summer in eastern North America, *P. lotor* feeds mainly on animal matter, including insects, earthworms, snails, bird's eggs, and small mammals. It also feeds on carrion and commonly visits creek edges to search for crayfish, frogs, fish, and other aquatic prey. During late summer, autumn, and winter, fleshy fruits and seeds, such as wild grapes, acorns, beechnuts, and berries, constitute the bulk of the raccoon's diet.

In the order Carnivora, notable omnivores are found in the families Procyonidae (raccoons, coatis, ringtails, cacomistles, kinkajous, and olingos), Herpestidae (mongooses), Eupleridae (Madagascar mongooses), and Viverridae (civets and genets).

Procyonids

The family Procyonidae (raccoons, coatis, and allies) includes 6 genera and 14 species, restricted to the New World; they typically inhabit forested temperate and tropical areas, usually near water. Body mass ranges from about 1 kg (2.2 lb) in the ringtail (*Bassariscus astutus*) and olingo (*Bassaricyon gabbii*) to 12 kg (26 lb) or more in raccoons. Procyonids typically have long, bushy tails (**prehensile** in the kinkajou, *Potos flavus*), with alternating light and dark rings, and obvious facial markings. They are **plantigrade** (walking on the soles of the feet); some have semiretractile claws, and all are adept at climbing trees. Dentition is generalized and adapted for an omnivorous diet, with fruit predominating in the kinkajou and olingo. The carnassials are fairly well developed only in the ringtail and cacomistle (*Bassariscus sumichrasti*). The ringtail (fig. 5.1), also called cacomistle, ringtail cat, miner's cat, rock cat, and civet cat, is the most colorful of the procyonids (Poglayen-Neuwall and Toweill 1988). Ringtails weigh 0.8 to 1.3 kg (1.7–2.8 lb) and resemble American martens in bodily contour. They sport a tawny-reddish coat and have a gray, foxlike face and a flashy, bushy tail banded with 14 to 16 black and white rings for its entire length. These **arboreal** carnivores inhabit rocky outcrops, foothills, pinyon-juniper woodlands, montane shrublands, and riparian woodlands from southwestern Oregon south to the state of Oaxaca, Mexico. Like margays, ringtails can rotate their hind foot 180°, permitting head-first descent of trees with great dexterity. Gene Trapp (1972: 552) studied ringtails in captivity and noted that "one young female walked upside down, slothlike, along a 0.5 centimeter [0.2 inches] cord strung between two supports, holding on the cord with the flexed digits of all four feet." Ringtails are well adapted for arboreal life, employing behaviors that include "chimney stemming" between vertical walls, "ricocheting" off vertical surfaces, and "power leaping" to reverse direction on narrow ledges. These **nocturnal** acrobats are notably omnivorous, with a diet of animal and plant material. They prefer grasshoppers, beetles, moths, and other insects, as well as spiders, small mammals, and fruits. Acorns, lizards, bird's eggs, and carrion are also consumed.

Herpestids

The family Herpestidae includes 14 genera and 33 species of mongooses. This Old World family is native to Africa, the Middle East, and Asia. Mon-

FIGURE 5.1. The ringtail (*Bassariscus astutus*), also called cacomistle, ringtail cat, and miner's cat, is the most colorful member of the raccoon family (Procyonidae). These arboreal carnivores inhabit rocky outcrops, foothills, pinyon-juniper woodlands, and montane shrublands from southwestern Oregon to the state of Oaxaca, Mexico. *Source:* J. G. Hall, ASM image library.

gooses generally are small; size ranges from the dwarf mongoose (*Helogale parvula*), with a body mass of 0.3 kg (0.6 lb), to the white-tailed mongoose (*Ichneumia albicauda*) at about 5.0 kg (11 lb). Herpestids occur in a variety of **terrestrial** and semiarboreal habitats. They are feeding **generalists** (feeding on a variety of foods), typified by a generalized dentition. Some species are solitary, whereas others are highly social and form groups. One such social species is the meerkats (*Suricata suricatta*), which form family groups in which individuals cooperate in activities such as rearing young and detecting predators. Meerkats have been extensively studied to test hypotheses about the evolutionary development of helping behavior by non-breeding individuals in social groups (O'Riain et al. 2000; T. J. Clutton-Brock et al. 2001). The small Asian mongoose (*Herpestes javanicus*) was introduced to the West Indies and Hawaii, where it is considered a pest. It preys on poultry and native fauna, especially nesting birds, and it carries rabies.

Viverrids

The civets and genets (family Viverridae) are found only in the Old World. The 15 genera and 35 very diverse species occupy tropical and subtropical habitats in Europe, Africa, the Middle East, and Asia. Civets and genets look like long-nosed cats with spots. They have long, slender bodies equipped with pointed ears and short legs. An exception is the binturong, which resembles a wolverine adorned with shaggy black hair, a long, thick prehensile tail, and a catlike head; binturongs are rather sinister looking but are apparently quite easy to tame—and they weigh too much to be considered small mammals. In physical appearance and life history, viverrids resemble members of other carnivore families and fill many equivalent niches. For example, some forms are **carnivorous** and parallel mustelids or felids in niche type. Others are omnivorous, such as procyonids, some are **frugivorous**, and others are scavengers, such as hyenas. Species may be **diurnal** or **nocturnal**, may occur singly or in groups, and may be terrestrial, semiaquatic, or arboreal.

Euplerids

The family Eupleridae comprises seven genera and eight species of Malagasy (Madagascar) mongooses. Species in this family were formerly categorized in two other families. The ring-tailed mongoose (*Galedia elegans*), the brown-tailed mongoose (*Salanoia concolor*), the narrow-striped mongoose (*Mungotictis decemlineata*), and the broad-striped and giant striped mongooses (*Galidictis fasciata* and *G. grandidieri*) were included in family Herpestidae. The Malagasy civet (*Fossa fossana*), fossa (*Crytoprocta ferox*), and falanouc (*Eupleres goudotii*) were placed in the Viverridae. Malagasy civets and mongooses range in size from the narrow-striped mongoose, weighing about 800 g (1.7 lb), to the arboreal fossa, at 20 kg (44 lb). They occupy habitats ranging from rainforests and open woodlands to savannahs and deserts. The diet of the smaller euplerids includes small mammals, birds, reptiles, eggs, insects, and fruits. The largest of the Malagasy carnivores is the arboreal catlike fossa; lemurs comprise more than one-half of its diet.

Mycophagy

Animals that consume fungi are referred to as **mycophagous** ("fungus-eating"). Fungi of various types are an important component in the diet of a diverse array of mammals, including "insectivores," herbivores, carnivores, and omnivores. Fogel and Trappe (1978) provide a thorough review of mycophagy in mammals, detailing both the specific taxa of fungi consumed and the mammalian groups known to exhibit mycophagy. Fungi preferred by mammals include the higher Basidiomycetes, Ascomycetes, and Phycomycetes (Endogonaceae) and lichens. These groups of fungi are especially well represented in the diets of squirrels, mice, spiny rats, and members of the **marsupial** families Potoroidae and Phalangeridae (Whitaker 1962, 1963a, 1963b; O. Williams and Finney 1964; Mangan and Adler 2002; Feldhamer et al. 2007).

It is noteworthy that 22 species of primates (including gorillas, bonobos, macaques, vervets, mangabeys, snub-nosed monkeys, marmosets, and lemurs) consume fungi. For most primates, less than 5% of their feeding time is allocated to mycophagy. However, some primates spend 12% to 95% of feeding time eating fungi—namely, buffy tufted-eared marmosets (12% of feeding time), Japanese macaques (14%), Goeldi's monkeys (29%), and two species of snub-nosed monkeys (95%).

For the mycophagous mammals, fungi constitute a principal component of the diet year-round, with seasonal peaks in consumption reflecting availability (Whitaker 1963a, 1963b). When ingested by mammals, sporocarps (the fungal structures that contain spores) pass through the digestive tract and are excreted without morphological change or loss of viability (Trappe and Maser 1976), while the other fungal tissues are digested. The rate of passage of spores varies from 12 hours in the Cascade golden-mantled ground squirrel (*Spermophilus saturatus*) to 24 hours in the deer mouse (*Peromyscus maniculatus*) (Cork and Kenagy 1989). Both **subterranean** (hypogeous) fungi and fungi growing under the bark of trees are important food sources for Eurasian red squirrels (*Sciurus vulgaris*) (Gurnell 1987). **Caching** of such fungi is common, and *S. vulgaris* may hang fungi on the branches of trees, next to the trunk, at heights of up to 8 m (26 ft); these fungal stores are typically short lived, with most fruiting bodies gone after two weeks (Lurz and South 1998).

Fleshy fungi are 70% to 90% water and provide the consumer with protein and phosphorus. Sporocarps of hypogeous fungi are reported to contain high concentrations of nitrogen, vitamins, and minerals. Fungi often contain complex carbohydrates associated with their cell walls, and thus many small mammals are unable to efficiently digest and access the nutritious fungal material. Some mammals have modifications of the digestive tract that enable them to use fungi as a primary food. Potoroids have an enlarged foregut in which fermentation of food is carried out by microbial symbionts. Long-nosed potoroos (*Potorous tridactylus*) can digest much of the cell wall and sporocarps of certain hypogeous fungi, thus accessing more of the fungi's available energy (Claridge and Cork 1994).

In certain coniferous and deciduous trees, the root systems are associated with the mycelia of certain types of fungi. The combination of host and fungus is called a **mycorrhiza**—a mutualistic relationship: the tree relies on filaments of the fungus to extract water and nutrients from the soil, and the fungus derives nutrition from the sugars produced by the tree (Trappe and Maser 1976). Mycorrhizal fungi do not produce aboveground fruiting bodies, and thus they rely on small mammals to disperse their spores. Many forest-dwelling small mammals, such as shrews, mice, voles, and squirrels, consume large quantities of spores of mycorrhizal fungi, which pass unchanged through their digestive tract. Detection of truffles (fungal fruiting bodies) by small mammals is primarily by olfaction. The dispersal of sporocarps by small mammals serves the essential function of reinoculating habitats for reestablishment of forests following natural catastrophes or deforestation.

CASE STUDIES
....

The Hero Shrew: Mysterious Insectivore

Hero shrews (*Scutisorex somereni*) inhabit swampy forests of northern Zaire, Rwanda, and Uganda (fig. 5.2A). These **insectivorous** shrews, also called armored shrews, were first described from a single specimen found in Uganda in 1913 by Oldfield Thomas. In 1917, J. A. Allen provided a thorough discussion of the unique skeletal characters of *Scutisorex*. He cited the field notes on *Scutisorex* furnished to him by Herbert Lang, leader of the American Museum of Natural History Congo Expedition (J. A. Allen 1917: 782). In his field notes, Lang provided details of the natural history and unique anatomy and behavior of hero shrews. The name "hero shrew" has its origin with the Mangbetu people of the Democratic Republic of the Congo. The Mangbetu and other natives of the region were well acquainted with the strength of these rather large shrews—weighing up to 90 g (3.1 oz). Lang noted that the natives believed that when setting out on dangerous adventures such as hunting elephants or engaging in warfare, they would be protected if they wore the charred body or heart of a hero shrew as a talisman. Carrying along some of the ashes of the hero shrew would transmit invincible qualities. Medicine derived from the hero shrew and prepared by

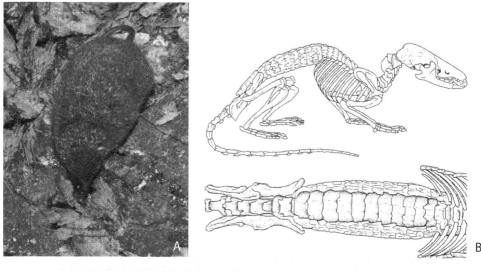

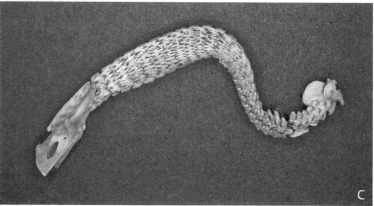

FIGURE 5.2. (A) The hero, or armored, shrew (*Scutisorex somereni*) inhabits forests of northern Zaire, Rwanda, and Uganda. (B) The *Scutisorex* skeleton and vertebral column, characterized by 11 lumbar vertebrae rather than the usual 5 or 6. (C) A *Scutisorex* vertebral column. The vertebrae are made particularly sturdy by the presence of many articular apophyses and interlocking spines, forming a basketlike structure not seen in any other mammal. *Sources:* (A) D. M. Cullinane, retouched by Carie Nixon, Illinois Natural History Survey; (B) S. Churchfield, *The Natural History of Shrews* (Cornell Univ. Press, 1990); (C) Howard P. Whidden.

the medicine men would provide protection from spears and arrows and act as a sort of passport to heroism.

Lang mentioned that local tribesmen would gladly show off the strength of the hero shrew by demonstrating its resistance to weight and pressure. One anecdote attests to the extraordinary strength of a hero shrew reportedly able to withstand the weight of a full-grown, 160 lb (72.5 kg) human balancing—barefoot and on one leg, no less—on the shrew. Things did not look good for the shrew, but once the tormentor jumped off, the shrew shivered a bit then scurried off, none the worse for this bizarre experience (J. A. Allen 1917: 782). It was crucial that during such a spectacle, the head of the shrew must be left free—the strength of the vertebral column coupled with the strong convex curve behind the shoulder evidently protected the heart and other organs from being crushed.

The hero shrew's spinal column has a morphology that is unparalleled among mammals (fig. 5.2B, C). It has 11 lumbar vertebrae (the typical mammalian number is 5 or 6) augmented by lateral, dorsal, and ventral interlocking spines, or apophyses, forming a unique basketlike structure. The ribs are robust, and the spine is four times more robust, relative to body mass, than that of any other vertebrate species (Cullinane, Aleper, and Bertram 1998). Further, each vertebra is very strong and sturdy, yet the vertebral column is quite flexible. The intervertebral joints are five times more resilient to axial torsion, per body mass unit, than those of the rat (*Rattus*) (Cullinane and Aleper 1998; Cullinane and Bertram 2000).

So what is the purpose or adaptive significance of this remarkable spinal morphology in the hero shrew? I doubt that this specialization evolved to cope with 160 lb humans jumping around the swamps of eastern Africa! Anatomists Sandy Whidden and Dennis Cullinane have examined in great detail the anatomy of the hero shrew's vertebral column and shoulder region, and they contend that there is no obvious functional significance that could help explain the animal's distinctive backbone. Some scientists speculate that this structural peculiarity may represent an adaptation for digging and burrowing to secure invertebrate food in the soil and leaf litter; however, this mechanism has not evolved in any other **fossorial** mammal. Kingdon (1974: 86) suggests that "the explanation for the peculiar backbone of *Scutisorex* is to be found in the need for a swamp-dwelling animal to get its body well clear of the ground." When compared with the streamlined and sleek posture of the typical crocidurine shrew, hero shrews seem to have a unique arched posture resulting from the inflexible lower back and displaced pelvis contrasting with the doubling up of the forequarters into an S-shaped bend. The long limbs, fingers, and toes acting together with this arched posture keep the body well above the ground (Kingdon 1997a: 145).

Dietary Nonconformists

Insects are a staple in the diets of many small mammals—namely, the bats, shrews, moles, numbats, hedgehogs, and Australian dasyurids. Insects are not generally the major food item for large mammals, with the exception of some xenarthrans that are **myrmecophagous** ("ant-eating"). For the most part, we would not expect insects to represent a staple in the diet of members of the orders Rodentia and Carnivora. However, rodents are somewhat omnivorous and labile in their choice of foods, and two insectivorous species of Carnivora, the aardwolf (*Proteles cristatus*) of Africa and sloth bear (*Melursus ursinus*) of the Indian subcontinent, although far too large to qualify as small mammals, are worthy of mention here. (Sometimes I have a problem drawing the line, because of my enthusiasm for such fascinating animals!)

Insects, such as several species of nasute harvester termites (*Trinervitermes*), form the chief food of the aardwolf of southern Africa (Koehler and Richardson 1990). A single aardwolf was estimated to consume about 105 million termites a year (Kruuk and Sands 1972). The sloth bear, weighing up to 150 kg (330 lb), relies heavily on a diet of termites, ants, and other insects for sustenance during most of the year. In Nepal, insects, especially termites, were found to compose more than 50% of the diet of *M. ursinus* from September through April (Joshi, Garshelis, and Smith 1997). Adaptations supporting the digging out and feeding on insects include long, curved claws, naked, long, protrusile lips for sucking, and the absence of the inner pair of upper **incisors**, resulting in a gap in the front teeth. These features are adaptive for digging up termite nests, sucking up the occupants, and blowing away debris in the fashion of a vacuum cleaner. Due to their impressively motile snout, sloth bears were once called lip bears (*Ursus labiatus*).

Another example of a small "carnivore" has gained recent fame on Animal Planet's popular television program *Meerkat Manor*: the slender-tailed meerkats (*S. suricatta*, order Carnivora) These small, gregarious mongooses inhabit thorn and grassland savannah of the Kalahari Desert of southern Africa. Many of us recall reading Rudyard Kipling's *Rikki-Tikki-Tavi*, with its account of the duel between the mongoose and the cobra, and probably deduced that mongooses feed mainly on snakes—but this is not true for all mongooses. For slender-tailed meerkats, 78% of their diet consists of insects—namely, larvae and adult Coleoptera—supplemented by small reptiles. The meerkat is a herpestid that forms family groups in which individuals cooperate in activities such as rearing young and detecting predators. As noted earlier in the chapter, meerkats have been extensively studied to test hypotheses of the evolutionary development of helping behavior by nonbreeders (O'Riain et al. 2000; T. H. Clutton-Brock et al. 2001). Also, meerkats give acoustically different alarm calls, depending on the type of predator encountered and the level of urgency of the threat (Manser 2001).

Another small carnivore dedicated to an insectivorous diet is the tiny bat-eared fox of eastern and southern Africa, weighing in at 2 to 5 kg (5–11 lb). These foxes consume primarily termites and beetles—close to 70% of their diet consists of harvester termites (*Hodotermes*) and dung beetles (family Scarabaeidae). A combination of extremely long ears and small, numerous teeth enhances the bat-eared fox's ability to detect, capture, and consume its prey.

Members of order Rodentia are notably omnivorous, but one member, the grasshopper mouse (*Onychomys*) of North America (fig. 5.3A), is unique among North American rodents in having a diet composed entirely of grasshoppers, crickets, and ground-dwelling beetles. Grasshopper mice inhabit semiarid grasslands, sand hills, and open semidesert shrublands of western North America. Morphologically, they have unusually long fingers and claws (*Onychomys* means "clawed mouse") that provide great dexterity in tasks such as grasping and manipulating their insect prey. These feisty mice have evolved specialized attack strategies to avoid the defensive secretions of insect prey such as beetles (*Elodes* and *Chlaenius*). When attacking a whip-scorpion, *Onychomys* first immobilizes the tail and then attacks the head. The grasshopper mouse is extremely territorial and highly aggressive toward intruders of the same sex. A subordinate mouse cornered in the wrong place is likely to be killed by the dominant individual with a series of bites through the neck at the base of the skull. I recall how I became aware that I had captured *Onychomys* in one of my aluminum Sherman live-traps. A combination of a very strong acrid odor and a sizable hole "bored" through the corner of the trap was a giveaway as to its former occupant.

A marsupial "equivalent" of the grasshopper mouse is the mulgara (*Dasycercus cristicauda*) (fig. 5.3B), a marsupial carnivore closely related to

FIGURE 5.3. (A) A member of order Rodentia (family Cricetidae), the northern grasshopper mouse (*Onychomys leucogaster*) is unique among rodents in having a diet composed entirely of grasshoppers, crickets, and ground-dwelling beetles. (B) The ecological equivalent to the North American grasshopper mouse is the mulgara (*Dasycercus cristicauda*), a marsupial carnivore that lives in the arid sandy deserts of Australia. Mulgaras are closely related to the Tasmanian devil (family Dasyuridae). *Sources:* (A) R. K. LaVal, ASM image library; (B) Gerhard Koertner.

the Tasmanian devil and the quolls that lives in arid, sandy deserts of central Australia. About the size of an eastern chipmunk (*Tamias striatus*), mulgaras specialize in consuming large insects, spiders, and rodents. They are reported to be relatively uncommon, but their numbers increase dramatically when prey such as house mice (*Mus musculus*) are abundant. Like *Onychomys,* mulgaras are a no-nonsense predator—but with etiquette.

They are reported to fiercely attack a mouse and devour it methodically from head to tail, inverting the skin in fastidious fashion as they dine.

In the western hemisphere, the southern limit to the family Soricidae occurs in the northernmost part of South America. Of the 10 or so species of shrews in northern South America, all are in the genus *Cryptotis*. These "small-eared shrews" live in the northwestern Andes Mountains, in Venezuela, Colombia, Ecuador, and Peru. The insect-eating niche filled by soricids in the north is occupied by a fascinating group of shrewlike marsupials in the south—the caenolestids (fig. 5.4A). This South American group of marsupials represents a distinct line of evolution that diverged from ancestral stock before the Australian forms diverged, and its members are now placed in their own family, the Caenolestidae, order Paucituberculata. The family Caenolestidae contains three genera and six species of "shrew" or "rat" opossums. They have disjunct distributions in South America. The dusky caenolestid (*Caenolestes fuliginosus*) and northern caenolestid (*C. convelatus*) live at high elevations of the Andes Mountains of western Venezuela, Colombia, and Ecuador. The gray-bellied caenolestid (*C. caniventer*) lives at lower elevations of southern Ecuador. The Andean caenolestid (*C. condorensis*) inhabits dense vegetation in cold, wet, high-elevation forests and meadows of Ecuador. Incan caenolestids (*Lestoros inca*) inhabit drier habitats of the Peruvian Andes, while long-nosed caenolestids (*Rhyncholestes raphanurus*) are denizens of the forests of southern Chile and adjacent Argentina, including Chiloe Island. Caenolestids are primarily nocturnal, insectivorous or omnivorous, and terrestrial. They have long snouts with numerous tactile whiskers, small beady eyes, and large ears, and the hind limbs are longer than the forelimbs. Senses of hearing and smell are well developed, but, like shrews, the caenolestids have poor vision. Adults weigh about 40 g (1.4 oz). Their total length is about 30 cm (11.7 inches), one-half of which is the long, fully haired, nonprehensile, ratlike tail. Females lack a pouch. Shrew opossums have 48 teeth. They are unique among New World marsupials in having a reduced number of incisors, the lower middle two of which are large and **procumbent**. They are voracious predators, subduing prey with their forcepslike incisors; the remaining lower incisors, the **canines**, and the first **premolars** are **unicuspid**. Caenolestids are nocturnal and forage on insects, earthworms, and other small invertebrate and small vertebrates in well-marked runways below the leaf litter. Caenolestids, plus several other groups of small mammals, have **incrassated** tails

FIGURE 5.4. The family Caenolestidae of South America contains the shrew or rat opossums, including (A) the long-snouted rat opossum (*Rhyncholestes raphanurus*). These diminutive mammals inhabit the northwestern Andes Mountains, in Venezuela, Columbia, Ecuador, and Peru. The insect-eating niche filled by shrews in the north is occupied by this fascinating group of shrewlike marsupials in the south. (B) Another South American group of small mammals that place insects high on their dietary hit list is the mouse opossums, such as *Marmosa robinsoni*. These voracious predators consume grasshoppers by inflicting bites to the head and thorax. *Sources:* (A) Peter L. Meserve; (B) J. E. Eisenberg, ASM image library.

that are used in storing fat for periods of **torpor**. I describe this fascinating adaptation (fat tails) in my discussion of **thermoregulation** in Chapter 8.

Another group of South American marsupials that place insects high on their hit list is the mouse opossums (fig. 5.4B). Widespread in the neotropics, mouse opossums inhabit many different ecological settings, ranging from **mesic** tropical forests to shrublands and banana plantations, from sea level to elevations of 3,400 m (11,200 ft). This group of didelphids (subfamily Didelphinae) includes the mouse opossums (*Marmosa,* 9 species), slender mouse opossums (*Marmosops,* 16 species), and woolly mouse opossums (*Microureus,* 6 species). *Thylamys* species (the "fat-tailed opossums") are also in this group. One of the smaller species, Linnaeus's mouse opossum (*Marmosa murina*), weighs between 13 and 44 g (0.46–1.5 oz); a more common, larger species is Robinson's mouse opossum (*Marmosa robinsoni*), weighing up to 110 g (3.8 oz). With prehensile tails, mouse opossums are at home in trees, where they make nests of twigs and leaves. They are generally solitary, hunting and nesting alone. Mouse opossums are nocturnal, feeding on insects, spiders, lizards, bird's eggs, chicks, and fruits. They are voracious predators; large grasshoppers are killed by several bites to the head and thorax.

Shrewlike mammals are also found in the Australasian family Dasyuridae (marsupial mice and Tasmanian devils). This family, home of the marsupial carnivores, is large and diverse, comprising 20 genera and 69 species—the second largest marsupial family in Australia, just behind the Macropodidae (kangaroos, wallabies, and pademelons). The carnivorous marsupials of Australasia range in size from the diminutive ningauis (*Ningaui*) and planigales (*Planigale*), ranging in body mass from 2 to 10 g (0.07–0.35 oz), to the Tasmanian devil (*Sarcophilus harrisii*), weighing in at about 9 kg (20 lb). In this book, of course, we are most interested in the small end of the family.

Vying with the **placental** Etruscan and pygmy shrews for title of the world's "smallest" terrestrial mammal are the tiny Pilbara ningauis, which weigh close to 2 g (0.07 oz); next in line are the planigales and the slightly larger dunnarts (*Sminthopsis*), ranging from 15 to 50 g (0.52–1.7 oz) (fig. 5.5). Ningauis inhabit dry grasslands and savannah, and some species seem to be adapted to arid conditions. They are primarily nocturnal, finding shelter in dense hummocks and hollow logs. Captive ningauis are known to undergo **daily torpor** in response to low temperature or food deprivation. The period of reproduction may extend from September to March (spring and summer). **Gestation** lasts 13 to 21 days. Usually, five to six young are car-

FIGURE 5.5. Many small marsupial carnivores inhabit diverse plant communities in Australia. These include (A) the diminutive southern ningaui (*Ningaui yvonneae*), averaging only 5 g (0.18 oz), and (B) the long-tailed dunnart (*Sminthopsis longicaudata*), distinguished by its very long, brush-tipped tail that acts as a counterbalance while foraging. *Sources:* (A) Fritz Geiser; (B) Bert & Babs Wells / Oxford Scientific.

ried in a simple pouch to the stage of weaning. Young remain attached to nipples until they reach about 42 to 44 days of age.

There are 21 species of dunnarts, or "narrow-footed marsupial mice," residing in many ecological communities in Australia. Some species, such as the common dunnart (*Sminthopsis murina*), show very high fecundity. Sexual maturity is achieved at 6 to 8 months of age, and females may produce two to three litters of up to 10 young on 8 to 10 teats. The breeding season may last for eight months; this is coupled with a 10- to 13-day gestation, repeated reproductions, reduced parental care, and a weaning period of 60 to 70 days. The dunnarts are a widespread and well-represented group inhabiting many different ecological settings, from eucalypt woodlands and savannahs to farmlands, arid grasslands, and deserts. The fat-tailed dunnart (*S. crassicaudata*) stores fat in its tail, which in well-nourished individuals may become carrot-shaped. The long-tailed dunnart (*S. longicaudata*) is distinguished from all other dunnarts by its very long, brush-tipped tail (more than twice the length of the head and body), which may act as a counterbalance while the animal forages in rocky outcrops of western Australia. During autumn and winter, dunnarts are reported to form communal huddles as a means of conserving body heat. They are also known to undergo torpidity as a means of combating cold.

Small dasyurids of Australia occupy all terrestrial and semiarboreal habitats, from deserts to high-elevation rainforests; these small carnivorous marsupials are voracious predators. Fisher and Dickman (1993) examined the foraging behavior of 21 species of Australian dasyurids, ranging from the tiny southern and Wongai ningauis averaging 5 g (0.17 oz) in body mass to the brush-tailed phascogale averaging about 160 g (5.6 oz). These are clearly no-nonsense predators. Unlike some of the placental soricids, dasyurids do not eat insects that they have not killed themselves; they rarely ingest large prey as carrion. Laboratory feeding experiments by Fisher and Dickman found that following capture of prey, dasyurids tend to orient large struggling prey by using their paws and will tread on the prey to expose the soft intersegmental cuticle. They inflict killing bites, then decapitate or dismember the prey by tearing and shredding the cuticle before pulling out the body contents. Dasyurids masticate the prey finely to facilitate release of nutrients from body fluids. The dentition of dasyurids totals 42 to 46 teeth and is specialized for their carnivorous or insectivorous diet—teeth are not designed for crushing; rather, the rows of **molars** act like a pair of sharp scissors, the upper molars shearing against the cusps of the lower to dissect the prey. Dasyurids weighing less than 16 g (0.56 oz) seem to derive their highest energy return by feeding preferentially on small prey, found by foraging along trails in microhabitats where small prey are most likely to be encountered.

Another fascinating dasyurid is the kultarr (*Antechinomys laniger*), an inhabitant of dry savannahs, grasslands, and deserts of inland Australia. Kultarrs are also called "jerboa-marsupials" because their body form, equipped with long hind limbs, is suggestive of the bipedal **saltatorial** gait typified by kangaroo rats of North America and jerboas of North Africa and the Russian steppe. However, recent studies employing high-speed cinematography have showed that kultarrs actually use a bounding gait, as do other dasyurids. Like other dasyurids, these desert-adapted carnivores consume invertebrates such as spiders, crickets, and cockroaches. Kultarrs are known to undergo daily torpor in response to low temperatures and withdrawal of food (Geiser and Baudinette 1987). During the breeding season, the crescent-shaped pouch becomes well developed, usually equipped with eight **mammae**. The period of **estrus** is about 35 days. Following a gestation period of about 12 days, females bear a litter of six young. Weaning occurs after 3 months, and young reach sexual maturity at 11.5 months. Unlike most dasyurids, the pouch of kultarrs does not open to the rear.

Of the small carnivorous dasyurids, species of the genus *Antechinus* are the most numerous and widespread. All 10 species of *Antechinus* inhabit forests throughout Australia and weigh less than 100 g (3.5 oz). Two species of *Phascogale* (red-tailed and brush-tailed phascogales) occur in woodlands of Australia. The brush-tailed phascogale (*P. tapoatafa*) is widespread in the coastal woodlands and is one of the most arboreal dasyurids. It was referred to by early settlers of Sydney as the "vampire marsupial" and "blood-thirsty killer"; actually, brush-tailed phascogales feed primarily on cockroaches, beetles, and centipedes. Several dasyurids, including the southern dibbler (*Parantechinus apicalis*), the red-tailed phascogale (*Phascogale calura*), and several species of dunnarts, are endangered because of reduced distribution and population size.

Gleaning, Hawking, Hovering, and Perch-Hunting

Most bats feed on insects taken on the wing; however, there are some remarkable **specialists**. Chiroptologists commonly separate bats' hunting techniques into several types, including gleaning, hawking, and perch-hunting;

however, keep in mind that these techniques are not mutually exclusive. As Norberg (1994) indicates, natural selection has favored specific wing shapes and designs best adapted for conserving energy and optimizing foraging success. Morphology, flight patterns, structure of **echolocation** calls, and feeding behavior are strongly linked; different bat species have evolved an optimal fit of such characters, tailored to a specific environment (Bogdanowicz, Fenton, and Daleszczyk 1999; Whitaker 2004). Some examples of foraging techniques and their practitioners are described here.

Gleaning bats typically fly slowly within the foliage and are able to carry heavy prey (Kalko and Schnitzler 1998). Low wing loading combined with low aspect ratios are adaptive for such aerodynamic abilities. Broad wings coupled with large tail membranes and rounded wing tips optimize maneuverability in dense vegetation. *Plecotus auritus*, the brown long-eared bat, belongs to the family Vespertilionidae and is widespread in Europe; it is a good example of a foliage gleaning bat that picks insects out of trees or off the ground. As one would expect from looking at its impressive ears, *Plecotus* forages mostly by passive listening—detecting prey by using vision rather than sonar. Typically, forest gleaners fly down and, with their large incisors and canines, grasp their prey off the surface of a leaf or trunk of a tree and move to a nearby feeding roost or temporary perch to eat it. Often they consume only the body of the prey, leaving the wings to fall to the ground. Gleaning is advantageous over aerial capture because moths, a favorite food of gleaners, cannot easily detect the bat's presence. Additionally, gleaners are less dependent on air temperature, and they can expand their food choice by catching nonflying prey such as coleopterans, trichopterans, dipterans, and spiders (Wilson 1997; Brack and Whitaker 2001).

Pallid bats (*Antrozous pallidus*) of the southwestern United States are gleaners that are impressive dive-bombers, consuming beetles, Jerusalem crickets, scorpions, beetles, and small lizards that they glean from the ground (fig. 5.6A) (Hermanson and O'Shea 1983). Golden-tipped bats (*Kerivoula papuensis*) of southeastern Australia feed by gleaning, flying slowly in dense vegetation and hovering and plucking orb spiders from their webs (Richards 1990; Van Dyck and Strahan 2008). As noted in Chapter 4 and further described below, certain species of bats are specialized and highly carnivorous, feeding on small vertebrates such as rodents, birds, frogs, lizards, small fish, and even other bats. Carnivorous bats are found in five families: Megadermatidae, Nycteridae, Noctilionidae, Vespertilionidae, and Phyllostomidae. The shape of the skull and teeth is a good indicator of diet. For ex-

FIGURE 5.6. Most bats feed on insects taken on the wing; however, there are some remarkable specialists. (A) The pallid bats (*Antrozous pallidus*) of the southwestern
United States are gleaners; they are reported to "dive-bomb" beetles, crickets, and
even small lizards. (B) Fringe-lipped bats (*Trachops cirrhosus*) of the family Phyllostomidae are perch-hunters—they remain on a perch until prey is detected, at which
point they fly to capture it. *Source:* Merlin D. Tuttle, Bat Conservation International.

ample, the carnivorous false vampire bat (*Vampyrum spectrum*), the largest bat in the New World, has a massive skull equipped with strong, sharp canine teeth and shearing molars adapted for crushing bones and cutting flesh. *Vampyrum* was once thought to be a true vampire bat, but it does not consume blood. Its diet consists of birds, bats, rodents, and some insects and fruits (Gardner 1977). Asian false vampire bats (*Megaderma lyra*) prey on small vertebrates, such as mice, chicks, and frogs, which are carried to the roost to be eaten. The fringe-lipped bat (*Trachops cirrhosus*), a phyllostomid, primarily consumes insects and small vertebrates such as frogs and lizards (fig. 5.6B) (Tuttle and Ryan 1981; Bruns, Burda, and Ryan 1989). Able to locate and distinguish between different species of frogs by listening for and analyzing their unique calls, *Trachops* can discriminate between poisonous and palatable species (Ryan and Tuttle 1983; Vaughan, Ryan, and Czaplewski 2000: fig. 1.64)

In aerial hawking, prey are pursued and caught in flight. Open space "hawkers" fly very rapidly and have large turning circles. Many sac-winged, free-tailed, and common bats fall into the hawking category—greater sac-winged bats (*Saccopteryx bilineata*), eastern red bats (*Lasiurus borealis*), and eastern pipistrelles (*Perimyotis subflavus*) forage by hawking. Aerial hawking has recently expanded to also exploit migrating songbirds. Bird-eating has been reported primarily in gleaning tropical bats, which occasionally capture resting birds (Norberg and Fenton 1988). However, Carlos Ibanez and his colleagues (2001), at the Donna Biological Station in Seville, Spain, discovered that migrating nocturnal songbirds are an important item in the diet of giant noctule bats (*Nyctalus lasiopterus*) from temperate regions of Spain and Italy (Dondini and Vergari 2000; Popa-Lisseanu et al. 2007). These bats, equipped with sharp canines and an efficient radar system, have the ability to prey on migrating birds on the wing. A joint effort by Spanish and Swiss scientists has confirmed aerial hawking in noctule bats, in studies using stable isotope tracers to track their diets. Recently, scientists in India and China have confirmed that great evening bats (*Ia io*) also consume songbirds on the wing (Thabah et al. 2007).

Perch-hunting, also called "hang-and-wait" hunting, is reported for many insectivorous and carnivorous gleaning bats of the Old World families of leaf-nosed bats (Megadermatidae) and slit-nosed bats (Nycteridae), as well as two neotropical species: d'Orbigny's round-eared bats (*Tonatia silvicola*) and the fringed-lipped bats (*Trachops cirrhosus*) of the family Phyllostomatidae (Kalko et al. 1999; Cramer, Willig, and Jones 2001). In perch-hunting, bats remain on a perch until a prey item is detected, at which point they fly to capture the prey. Perchers that consume nonflying prey include false vampire bats (*Megaderma lyra, M. spasma, Macroderma gias,* and *Cardioderma cor*) of the family Megadermatidae (Vaughan 1976). Perch-hunters are characterized by short, broad wings that adapt them for excellent maneuverability in dense forest environments. *T. cirrhosus* is a very efficient predator of frogs of the neotropics; it locates breeding frogs by keying into their mating calls and plucks them from the land or surface of the water, flying low and grabbing them with their well-developed teeth. Fringed-lipped bats employ echolocation as well as listening to acoustic cues of frogs. They can discriminate between poisonous and palatable frogs on the basis of acoustic cues alone (Ryan and Tuttle 1983).

Many **nectarivorous** bats hover when feeding, although many sit on the flower when feeding. As described in Chapter 3, nectarivorous bats include New World bats of the family Phyllostomidae, such as Mexican long-nosed bats (*Leptonycteris nivalis*), Mexican long-tongued bats (*Choeronycteris mexicana*), and tube-nosed nectar bats (*Anoura fistulata*). Parallel adaptations for nectarivory are also known in megachiropterans such as the long-nosed fruit bats (*Macroglossus*) and tiny blossom bats (*Syconycteris*). Although both the New and Old World bats feed primarily on nectar and pollen, their diets also include fruits and insects.

Like other mammals, bats have well-differentiated teeth, including incisors, canines, premolars, and molars, and the teeth are deciduous: a set of milk teeth is replaced by permanent teeth early in development. The milk teeth of bats are unique. They have pointed hooks that permit the young to attach firmly to the mother's nipples, essential for the pups to hold on while mom is flying! Small insectivorous bats have 38 teeth. The molars of insectivorous bats are characterized by W-shaped cusps adapted for crushing exoskeletons. In contrast, vampire bats have only 20 teeth, as their meal of blood does not require chewing.

PART TWO

Environmental Adaptations

The success of small mammals derives from their sophistication in form, function, and behavior. Small mammals are widely distributed throughout the earth, in environments ranging from tundra to deserts, each characterized by different extremes of climate. As a result of such strong selection pressures, small mammals have developed unique suites of adaptations to enhance survival. These adaptive strategies are diverse and highly sophisticated, with variation at the levels of ecosystems, species, and populations of individuals.

In this part of the book, I introduce some basic terminology of the physiology of environmental adaptation: endothermy, homeothermy, euthermy, heterothermy, and poikilothermy, as well as ambiguous terms such as warm-blooded and cold-blooded. Most mammals maintain their body temperature between 36°C and 38°C (96.8°F–100.4°F). To do this, they must balance heat gains and heat loss against such forces as radiation, conduction, convection, and evaporation. Each species has unique coping mechanisms and peculiarities of its thermoneutral zone, thermal conductance, and metabolic rate. These differences among species help determine their distribution. It is also interesting to assess how different species of small mammals conform to some long-standing "ecogeographic rules."

Unlike many birds species in cold regions, small mammals do not undertake long migrations and must remain as residents during winter months. As a result, over their evolutionary history, they have developed mechanisms of avoidance of and resistance to cold. Avoidance includes behavioral mechanisms (communal nesting, nest construction, food hoarding, foraging dynamics, and reduced levels of activity) and anatomical and physiological mechanisms (reduced body mass in winter, decreased thermal conductance, and heterothermy). Small mammals resist cold by increasing thermogenic capacity with higher basal metabolic rate and shivering and nonshivering thermogenesis.

The challenges confronting small mammals in the hot, dry conditions of desert environments are even more severe. Water is essential for survival, so water economy and the anatomy and physiology of the kidney play a crucial role in desert animals. Small mammals of desert environments conserve water by eating succulent plants and using metabolic water. Their adaptive strategies include unique patterns of insulation, metabolic profiles, and behavioral ways of dealing with heat stress.

CHAPTER SIX

Endothermy

Land mammals inhabit environments characterized by extremes in temperature, precipitation, and altitude. Marine environments also vary greatly in temperature and in water pressure at different depths. Species of **terrestrial** mammals may be faced with temperatures ranging from near −65°C (−85°F) in the Arctic to 55°C (131°F) in Death Valley, California. Daily fluctuations during winter can be extreme: in Montana, the temperature may range from 6°C to as low as −49°C (−86°F to −56.2°F) during one day. Summer is no relief for mammals, and a hot spell in western Australia may reach 38°C (100.4°F) and last for 162 continuous days! Marine mammals also experience temperature variation: water temperatures range from −2°C (28.4°F) near the north and south poles to 30°C (86°F) near the equator.

Most mammals do not undertake long migrations; they remain as residents during winter months, adapted over the course of evolution through mechanisms of *avoidance* of and *resistance* to cold (table 6.1). Avoidance of cold entails energy conservation, whereas resistance requires energy expenditure—a debt that must be repaid. Mammals rarely rely on a single mechanism to enhance survival in the cold, instead exhibiting a suite of strategies that integrate behavioral, anatomical, and physiological specializations finely tuned to a specific habitat and lifestyle (Merritt and Zegers 2002). Thermoregulatory mechanisms vary according to taxon and the attributes of each species.

To survive in deserts, mammals must cope with a variety of demanding environmental challenges, such as intense heat during the day, cold nights, a paucity of water and cover, and a highly variable food supply. Desert ecosystems are widespread and abundant—35% of the earth is covered with deserts. Mammals have successfully colonized desert ecosystems, as evidenced by the rich and diverse fauna of heteromyid and dipodid rodents

TABLE 6.1. *Winter Survival Mechanisms of Small Mammals*

Avoidance

    Body size
    Insulation
    Appendages
    Coloration
    Modification of microclimatic conditions
        Communal nesting
        Construction of elaborate nests
    Foraging zones
    Food hoarding
    Reduction in level of activity
    Reduction in body mass
    Heterothermy

Resistance

    Increase in thermogenic capacity through increase in basal metabolic rate, nonshivering
        thermogenesis, shivering

Source: Carie Nixon, Illinois Natural History Survey.

in the deserts of North America and the Old World, respectively. Allan Degen (1997) provides an excellent review of studies examining the **adaptations** of mammals to desert environments.

The challenges faced by mammals in desert regions are even more severe than those encountered in cold regions; water as well as food is scarce, and the problem in temperature regulation is reversed. In some deserts, mammals may have to cope with air temperatures that reach 55°C (131°F) and ground temperatures that exceed 70°C (158°F). Unlike arctic mammals concerned with conserving heat, desert mammals must dissipate heat or avoid it to maintain **euthermy.**

Seed-eating rodents living in desert ecosystems tend to have low basal **metabolic rates.** Examples include heteromyids of North America and murids of Australia and Asia. Alan French (1993) summarizes the factors related to energy consumption in heteromyid rodents. They may use low metabolic rates to reduce overheating when they occupy a closed burrow system or to reduce pulmonary water loss in a dry environment. Metabolic rates of heteromyids are about one-third lower than those of other mammals when at rest. In addition, metabolic reductions are possible when the mammals undergo a form of heterothermy called **estivation** during times of food scarcity.

The properties of a mammal that influence the exchange of energy are metabolic rate, rate of moisture loss, thermal conductance of fat or fur, ab-

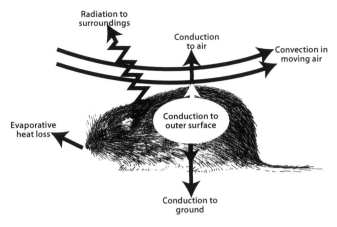

FIGURE 6.1. Because of the high surface-to-volume ratio of small mammals, they are constantly coping with heat loss from the environment through conduction, convection, and, in some cases, radiation. *Source:* Carie Nixon, Illinois Natural History Survey, with prairie vole drawn by Aleta Holt, in J. E. Hofmann, *Field Manual of Illinois Mammals* (Illinois Natural History Survey, 2008).

sorptivity to radiation, and the size, shape, and orientation of the body (fig. 6.1). Most of the heat produced by mammals is lost to the environment passively by *radiation, conduction,* and *convection* (air movement) to the cooler surroundings and by *evaporation* of water. A mammal's **heat load** (sum of environmental and metabolic heat gain) is roughly proportional to its body surface area. Because small mammals have a much larger surface-area-to-volume ratio, they lose heat more readily than large mammals. Using the surface-to-volume relationship, we can estimate the quantity of water required to dissipate a certain heat load. A small mammal such as a kangaroo rat would need to lose (by evaporation) a great deal more water (relative to body size) to eliminate its heat load than would a large mammal. To avoid this, small mammals escape the heat (curtail the heat load) by retreating to underground burrows during the day. In addition, desert mammals can decrease their body temperature to conserve water and decrease heat load.

Maintaining an internally regulated body temperature—**endothermy**—offers numerous benefits for mammals coping with extreme climates. Both mammals and birds have high metabolic rates—at least eight times those of ectotherms (in **ectothermic** animals, body temperature varies with environmental temperature). In terms of energy expenditure, maintaining high body temperatures in addition to high levels of activity is costly, but it has the advantage of enhancing coordination of biochemical systems, increasing information processing, and speeding central nervous system functions. As a result, mammals have a refined neuromuscular system, thus enhancing their efficiency at capturing prey and escaping from predators. Mammals gain independence from temperature extremes in nature, can extend activity periods over a 24-hour period, and colonize many environments and ecological niches throughout the world. Mammals can match their thermoregulatory pattern to suit a given environment and thus take advantage of nutritional resources year-round.

The evolution of endothermy in mammals has been the subject of considerable debate (Grigg, Beard, and Augee 2004; Kemp 2006). Endothermy is thought to have evolved from an ectothermal condition two or three times in the Mesozoic Era. Internal heat production may have evolved in the late Triassic in the group of mammal-like reptiles, in response to selective pressure favoring sustained activity and temperature regulation.

To keep a constant body temperature, mammals must maintain a delicate balance between heat production (energy in) and heat loss (energy out) (Feldhamer et al. 2007). Heat is produced through **metabolism** of food or fat, cellular metabolism, and muscular contraction. Factors that influence the exchange of energy between a mammal and the environment are sunlight (solar radiation), reflected light, thermal radiation, air temperature and movement, and the pressure of water vapor in the air. The properties of a mammal that influence the exchange of energy are metabolic rate, rate of moisture loss, thermal conductance of fat or fur, absorptivity to radiation, and the size, shape, and orientation of the body. As mentioned earlier, most of the heat produced by mammals is lost to the surroundings passively by radiation, conduction, and convection (air movement) to a cooler environment and by the evaporation of water. Mammals must adjust their energy balance to meet different demands in their environment (see Feldhamer et al. 2007: fig. 9.1).

Mammals regulate body temperature by continuously monitoring outside temperature at two locations: on the surface of the skin and at the hypothalamus. The hypothalamus, or mammalian "thermostat," is located in the forebrain, below the cerebrum, and operates by comparing a change in body temperature with a reference temperature, the **set point**. Each species may have a different set point, or comfort zone, which is set at the hypothalamus. For most eutherian mammals, heat is generated by muscle contraction (i.e., shivering), **brown adipose tissue,** also called brown fat (by **nonshivering thermogenesis**), and activity of the thyroid gland. In **marsupials** and monotremes, heat production is due primarily to the activity of skeletal muscle (Nicol and Andersen 1993). For mammals inhabiting variable environments, body temperature is maintained around their set point, with each mammal's set point determined by such factors as insulation, behavioral postures, activity levels, and microclimatic conditions.

Most eutherian mammals maintain a core body temperature of about 38°C (100.4°F). Monotremes and marsupials tend to have lower body temperatures, with echidnas and platypuses having normal body temperatures ranging from 28°C to 33°C (82.4°F–91.4°F) (Grigg, Beard, and Augee 1989; Grigg et al. 1992). Each species has a range of environmental temperatures, referred to as the **thermoneutral zone,** within which the metabolic rate is minimal and does not change as ambient temperature increases or decreases. The upper and lower limits of the thermoneutral zone are referred to as the upper and lower "critical temperatures"(fig. 6.2A). Decreasing environmental temperatures require that an animal increase its metabolic rate to balance heat loss. The temperature at which this becomes necessary, the **lower critical temperature,** varies from species to species and is seasonally adjusted by the interplay of insulatory thickness, behavioral attributes, and integration with the hypothalamus. As environmental temperatures decrease, adjustments in **thermal conductance** (the rate at which heat is lost from the skin to the outside environment) and metabolism are necessary if a species is to maintain euthermy. If this is not possible, death due to **hypothermia** (low body temperature) results.

At the other end of the thermoneutral zone is the **upper critical temperature,** the temperature above which animals must dissipate heat by **evaporative cooling** to maintain a stable internal temperature. This upper limit is important to mammals living in desert environments and is less variable than the lower critical temperature. Mammals employ many different mechanisms— such as seeking shelter in cooler, underground burrows and restricting surface activity to hours of darkness—to avoid upper critical temperatures. Furthermore, evaporative cooling can be facilitated by pulmonary means, **panting,** sweating, spreading saliva, or estivation. If a mammal cannot combat an increase in temperature by dissipating heat through evaporation or by the muscular activity of panting, its body temperature will keep increasing, ultimately resulting in death.

In euthermic animals, heat production must equal heat loss, and heat

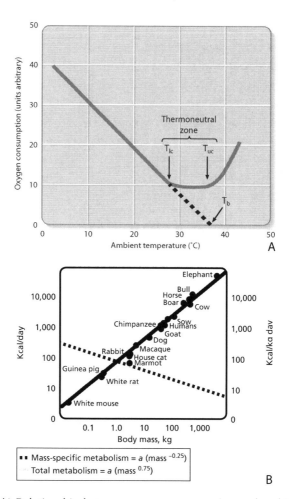

FIGURE 6.2. (A) Relationship between oxygen consumption and ambient temperature (°C) in a hypothetical mammal. T_{lc} = lower critical temperature; T_{uc} = upper critical temperature; T_b = core body temperature. (B) A comparison of body size and metabolism. The figure compares the log of body size (kg) in placental mammals with the log of total metabolism (kcal/day; solid line) and mass-specific metabolism (kcal/kg per day; broken line). a = proportionality constant (3.8) for placental mammals. *Sources:* (A) Feldhamer et al. 2007, data from M. Kleiber (*Hilgardia* 6:315–353, 1932); (B) Feldhamer et al., data from G. A. Bartholomew, in *Animal Physiology: Principals and Adaptations* (4th ed., ed. M. S. Gordon et al., Macmillan, 1982).

loss is proportional to surface area. Resting, or basal, metabolic rate is the volume of oxygen consumed per unit of time at standard temperature and pressure and is known for many species of mammals. It makes intuitive sense that larger animals require more food to maintain their body temperature and level of activity. As shown in figure 6.2B (solid line), larger animals consume more oxygen (they eat more food) than smaller animals—this is **total metabolic rate**. To compare oxygen consumption between animals of different size, metabolic rate is adjusted for body size and expressed as **mass-specific metabolic rate** (rate of oxygen consumption per gram of body mass); this decreases as body size increases (fig. 6.2B, broken line). Given this relationship, we can estimate metabolic rate from the body mass of a species. The relationship of mass-specific and whole-animal metabolism is complex, and authors do not always agree on the interpretation of this relationship (Packard and Boardman 1988; Hays and Shonkwiler 1996; Hays 2001; McNab 2002). The slope of the solid line in figure 6.2 has a value of about 0.75; that is, the metabolic rate scales to body mass to the three-fourths power; this relationship is generally known as Kleiber's law (Kleiber 1932, 1961; Feldhamer et al 2007).

CHAPTER SEVEN

Heterothermy

....

The condition of **hypothermia** is familiar to humans. During exposure to cold, our internal mechanisms may be unable to replenish the heat that is lost to the surroundings. Our body temperature drops below that required for normal **metabolism**, we begin to shiver, and bodily functions are compromised. Many other mammals and some birds employ a mechanism of controlled **heterothermy**, which also entails a decline of body temperature but, in contrast to hypothermia, involves the ability to relax **homeothermic** (temperature-maintaining) responses, thus permitting the core body temperature to fall—even to levels closely approximating low ambient temperatures (R. W. Hill and Wyse 1989: 130).

Torpor

....

In some types of heterothermy, the mammal remains quite alert even at low body temperature. For example, short-beaked echidnas (*Tachyglossus aculeatus*) can be active and foraging normally with a body temperature of 20°C (68°F), which is 12°C below normal (Brice et al. 2002; Kuchel 2003). However, in many cases, a low body temperature is accompanied by a strongly depressed **metabolic rate** (MR), respiration rate, and heart rate. The term **torpor** is used for these relatively low body temperatures when the animal is inactive and exhibits reduced responsiveness. Despite the decreased body temperature, torpor is not a state in which **thermoregulation** is simply abandoned. Rather, it involves regulation of body temperature at a new level, with a new **lower critical temperature**. If ambient temperature falls below this new critical minimum, MR will be increased to maintain the critical body temperature.

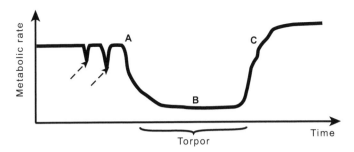

FIGURE 7.1. The pattern of changes in metabolic rate during entry into and exit from torpor in a small mammal. Torpor consists of three phases: *A*, the entry phase, characterized by declining body temperature (dashed arrows show preliminary "trial runs" when metabolism transiently slows); *B*, the middle phase, characterized by low and stable body temperature, metabolic rate, respiration rate, and circulation (during this time the mammal is relatively inactive and unresponsive); and *C*, the final phase, typified by increasing body temperature, metabolic rate, respiration rate, and circulation. The increase in body temperature in the final phase may be achieved through internal heat production or passive rewarming. *Source:* Carie Nixon, Illinois Natural History Survey, adapted from P. Willmer, G. Stone, and I. Johnston, *Environmental Physiology of Animals* (Blackwell Science, 2000).

Torpor consists of three phases (fig. 7.1):

1. The *entry phase* is usually characterized by a reduction in MR followed by declining body temperature, which in turn further reduces MR. This phase sometimes begins as a short series of transient drops in body temperature and return to **euthermy** (normal body temperature) followed by a single smooth, rapid drop to the new low body temperature.

2. The *middle phase*, the bout of torpor itself, is characterized by low and stable body temperature, MR, respiration rate, and circulation. During this time the mammal is relatively inactive and unresponsive.

3. The *final phase* is typified by increasing body temperature, MR, respiration rate, and circulation. The increase in body temperature may be achieved through internal heat production or through passive rewarming—that is, basking in the sun (Geiser 2004).

Bouts of torpor are employed in a variety of ways by a diverse group of mammals living in environments ranging from arctic to tropical. There are three major strategies in which mammals employ torpor: estivation, daily torpor, and hibernation. The terminology for torpor can be quite confusing and has been used inconsistently (M. S. Gordon et al 1982; Geiser and Ruf 1995). Estivation is the least physiologically distinct of the three strategies and has not been as intensely studied as either hibernation or daily torpor. **Estivation** involves a sequence of prolonged bouts of torpor during the relatively high air temperatures of summer. It is often associated with hot, dry conditions; it is used by some **marsupials** and "insectivores" but is most common among rodents in **xeric** environments. Estivation is employed by various members of the family Heteromyidae. This family includes pocket mice (*Perognathus, Chaetodipus*), spiny pocket mice (*Heteromys, Liomys*), kangaroo rats (*Dipodomys*), and kangaroo mice (*Microdipodops*) (French 1993; Feldhamer et al. 2007). Most members of this New World family are found in arid regions; many species are residents of the dry, desert conditions of western North America, whereas spiny pocket mice occupy wetter, tropical habitats of Central and South America.

Kangaroo mice (body mass 10–14 g, or 0.35–0.5 oz) are found in the Upper Sonoran sagebrush desert of western North America. Seeds are a staple in their diet and are gleaned from the sand and hoarded in underground burrows. Although environmental temperatures may range from 0°C to 30°C (32°F–86°F), kangaroo mice rarely experience drastic changes in ambient temperature, given their **fossorial** (burrow-digging) and **nocturnal** habits. In most deserts, seed production is seasonal and limited to very brief periods. As a result, the constant maintenance of normal body temperature would be a waste of energy for **granivorous** rodents such as kangaroo mice. The duration and frequency of estivation by kangaroo mice is contingent on the availability of seeds and environmental temperature. For example, when food is in excess and environmental temperatures are high, kangaroo mice exhibit normal body temperature; if food is limited and temperatures unfavorable, they undergo estivation. Periods of estivation may last up to several consecutive days. French (1993) and Degen (1997) provide interesting and thorough discussions of strategies of torpor in the family Heteromyidae.

Daily torpor is typified by bouts of torpor limited to a duration of less than 24 hours. This strategy has been documented in mammals from five of the six zoogeographic zones of the world, with practitioners of daily torpor residing at all latitudes (Lovegrove 2000). Small mammals may employ daily torpor in various seasons of the year.

While many species of small tropical bats seem to use torpor as a last resort, linked with energy deficits, many species of plain-nosed bats (Vespertilionidae) and horseshoe bats (Rhinolophidae) frequently use daily torpor in spring, summer, and autumn. However, during the reproductive season in spring and early summer, male and female bats are known to adopt different thermoregulatory behaviors: females normally maintain high body temperature throughout pregnancy and **lactation**, while males undergo bouts of daily torpor (Kurta, Johnson, and Kunz 1987; J. L. Hamilton and Barclay 1995; Dietz and Kalko 2006).

During winter, while most bats in temperate regions of the world are hibernating, striped skunks undergo bouts of daily torpor, with body temperature reaching 26°C (78.8°F)—the lowest torpid body temperature recorded for any carnivore (Hwang, Larivière, and Messier 2007). Mice of the family Cricetidae commonly survive winter by employing daily torpor coupled with communal nesting. In north temperate climates, mice of the genus *Peromyscus* may be active on a warm January day, but when ambient temperatures reach 2°C to 5°C (35.6°F–41°F), these small rodents may rapidly decrease body temperature to 13°C (55.4°F) and undergo bouts of torpor that last for less than 12 hours. Most hamsters do not exhibit torpor; however, Siberian hamsters (*Phodopus sungorus*) undergo daily torpor during winter, with a body temperature near 19°C (66.2°F). Color change and torpor in these hamsters are variable; some individuals turn white in winter and exhibit torpor, whereas others remain brown and do not enter torpor.

Species of shrews in the eastern hemisphere, the white-toothed shrews (subfamily Crocidurinae), evolved in warmer, more southern latitudes, and some have the ability to enter daily torpor (Nagel 1977). These include *Crocidura russula, C. suaveolen, Diplomesodon pulchellum,* and even the smallest of all **terrestrial** mammals, the Etruscan shrew (*Suncus etruscus*). Species of shrews in the western hemisphere, the red-toothed shrews (subfamily Soricinae), evolved in more northerly latitudes and are not known to enter torpor—most likely related to their comparatively high metabolic rates (Churchfield 1990). Peter Vogel, Michel Genoud, and coworkers of the University of Lausanne in Switzerland have conducted some interesting work on thermoregulation and behavior of white-toothed shrews (*C. russula* and *S. etruscus*) (Vogel 1976, 1980). They estimate that torpor results in a savings of approximately 15% of daily energy expenditure in greater white-toothed and Etruscan shrews (Genoud 1988). It seems that these shrews use torpor as a short-term strategy in dealing with emergencies in the form of food shortages.

Hibernation

Winter is often characterized by a decline in environmental temperature, of course, and for many mammals it is also a season when food supply dwindles. The combination of these factors makes the defense of an elevated body temperature problematic. **Hibernation** is a strategy used by some species of small mammals to survive the long, cold winter months of food deprivation. Hibernation is a pattern of torpor that occurs throughout the winter season and is characterized by a series of prolonged bouts of torpor in which body temperature is usually very close to environmental temperature. In fact, the body temperature of a hibernating mammal may be within a few degrees of freezing point, representing a decline in body temperature of more than 32°C (58°F). The duration of the individual bouts of torpor during hibernation may range from 96 to 1,080 hours (Geiser and Ruf 1995; Geiser 2004). A surprisingly broad spectrum of mammals employ hibernation, with representatives from all three subclasses in the class Mammalia—including the egg-laying echidnas, the tiny marsupial pygmy possums, many rodents such as ground squirrels and dormice, some species of hedgehogs, and even a small primate, the fat-tailed dwarf lemur (*Cheirogaleus medius*) (Dausmann et al. 2004). Contrary to popular belief, bears are not among the mammals that hibernate; rather, these large members of the order Carnivora undergo a period of **winter lethargy**, which involves a decrease in body temperature of only 5°C to 7°C below their active level of 38°C (Harlow et al. 2004). If the grizzly bear were to hibernate, the task of increasing the temperature of its 386 kg body from 5°C to 38°C would require an astounding 11,116,800 calories! Because of this extraordinary amount of energy necessary to increase a very low body temperature of a large body, hibernation is not possible or even necessary in very large mammals, which can store sufficient energy in the form of internal fat to meet winter demands. The largest mammals to undergo hibernation are the marmots (*Marmota*), which weigh about 5 kg (11 lb).

For some species, the environmental cues that signal preparation for hibernation are associated with the time of year and involve a combination of low temperature and lack of food. For others, entry into hibernation may occur without an external stimulus. Whatever the case, many hibernating mammals accumulate body fat prior to the hibernation season. In the autumn, the body mass of the woodchuck of North America (*Marmota monax*) is approximately 30% greater than in early summer; the increase in body mass before hibernation reaches 80% in golden-mantled ground squirrels (*Spermophilus lateralis*). This ability to accumulate body fat has even given one hibernator, the dormouse, a reputation as a gourmet meal. Dormice occur throughout Europe, Africa, Asia, and southern Japan, where they inhabit forests, shrublands, residential areas, and rocky outcrops (fig. 7.2). They may shelter in tree cavities, attics, tree roots, piles of mulch, or even **subterranean** burrows. These semiarboreal rodents are **omnivores** whose diet varies according to season; furthermore, diet varies among species according to the region they inhabit. Some species are known to consume pears, plums, apples, and grapes and are reported to have destroyed one-third of the grape crop in northern Caucasus. African dormice are the most **carnivorous**, preying on spiders, earthworms, and small vertebrates, as well as eating eggs and fruits. Dormice have a simple stomach and are the only rodent that lacks a **cecum**, suggesting a diet low in cellulose.

The fat, or edible, dormouse (*Glis glis*) of Europe is the largest member of the family (body mass 200 g, or 7 oz). Ancient Romans considered it a great table delicacy. Colonies of dormice were maintained in large enclosures planted with nut-producing shrubs and provided with nesting sites. In preparation for a feast, dormice were confined to large earthenware jars called "gliraria" (the dormice's family name is Gliridae), where they were fattened on chestnuts and acorns. Dormice lived at the bottom of the jar in a thick layer of oily seeds. After several months they became very fat and were served up to gourmet diners as a special treat. In Slovenia since the thirteenth century, dormice have been considered highly suitable for human consumption, hence their alternative name, "edible dormice." This dormouse is still trapped for its fat and for use as food.

The hibernation period of dormice may last up to nine months, with the precise length of dormancy varying with species and geographic region. In addition to hibernation, dormice are also known to make use of other forms of torpor: daily torpor and estivation (Wilz and Heldmaier 2000). These frequent periods of inactivity throughout the year certainly lend credence to their reputation for sleeping, as evidenced in the French phrase *dormir comme un loir,* "to sleep like a dormouse," equivalent to the English

FIGURE 7.2. The hazel dormouse (*Muscardinus avellanarius*) undergoes true hibernation, which may last for up to nine months. *Source:* Rimvydas Juskaitis.

expression "to sleep like a log." And of course, Lewis Carroll's *Alice's Adventures in Wonderland* featured a sleepy dormouse at the Mad Hatter's tea party (fig. 7.3):

> There was a table set out under a tree in front of the house, and the March Hare and the Hatter were having tea at it: a Dormouse was sitting between them, fast asleep, and the other two were using it as a cushion, resting their elbows on it, and talking over its head. "Very uncomfortable for the Dormouse," thought Alice; "only, as it's asleep, I suppose it doesn't mind."

Australia's lone alpine marsupial hibernator, the mountain pygmy possum (*Burramys parvus*) (fig. 7.4A), prepares for hibernation primarily by feasting on a single species of insect, the moth *Agrotis infusa* (Geiser et al 1990). Seeds, berries, flowers, and fruits may play a role in the diet as well, but during summer the moths are plentiful and dormant in crevices of boulder banks. The fat bodies of the insect provide the pygmy possums with an excellent source of protein and energy-rich food. The possums feast on the moths until the body mass of the small marsupial has doubled from about 40 to 80 g prior to winter.

THE MAD TEA-PARTY.

FIGURE 7.3. The dormouse's reputation of a long period of dormancy and a sleepy nature is featured in Lewis Carroll's *Alice's Adventures in Wonderland* at the Mad Hatter's Tea party. *Source:* Sir John Tenniel.

Mountain pygmy possums are the only Australian mammal limited in distribution to alpine and subalpine regions characterized by snow cover for up to six months. During the winter months of April to October, these possums hibernate in nests situated in boulders beneath a deep mantle of snow. In winter, their body mass declines as they mobilize stored fat. Most of this mobilization of body fat is due to the energetically demanding process of increasing a nearly freezing body temperature to euthermy. Every species of mammal that hibernates experiences the process of increasing and decreasing body temperature multiple times throughout the period of hibernation. Gerhard Körtner and Fritz Geiser (1998) of the University of New England, Armidale, Australia, used temperature-sensitive radio-collars to monitor the body temperature of pygmy possums. They found that the hibernation season of the mountain pygmy possum, like all other hibernators, is interrupted by frequent periods when body temperature rises to euthermy and then falls back into a deep bout of torpor (fig. 7.4B).

The mountain pygmy possum was first described in 1895 from a fossil found in Wombeyan Caves, New South Wales, and for decades was thought to be extinct. Approximately 70 years later, in August 1966, a *live* pygmy possum was found by chance in a ski hut on Mount Hotham in the Snowy Mountains, Victoria (Mansergh and Broome 1994). This great find was immediately telegraphed to W. D. L. Ride by R. M. Warneke: "Burramys ex-

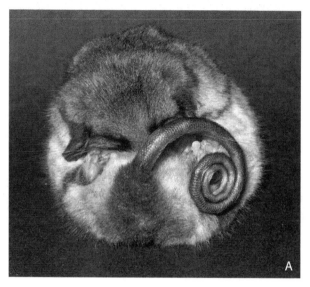

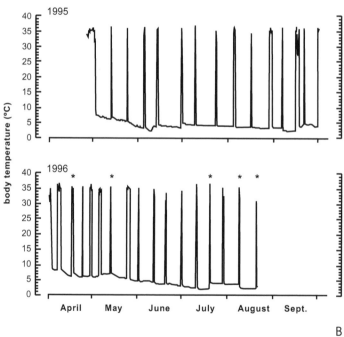

FIGURE 7.4. (A) The mountain pygmy possum (*Burramys parvus*) is the only alpine marsupial hibernator of Australia. (B) Hibernation patterns of a female mountain pygmy possum (*B. parvus*) in 1995 and 1996. Asterisks indicate a change of hibernaculum. *Sources:* (A) Fritz Geiser; (B) Carie Nixon, Illinois Natural History Survey, adapted from G. Körtner and F. Geiser (*Oecologia* 113:170–178, 1998).

tant STOP not repeat not extinct STOP live male captured Mount Hotham STOP am trying for female" The live pygmy possum was handed over to the Fisheries and Wildlife Department in Victoria, where it was identified. It was maintained in captivity, its captors hoping to find a female. However, in May 1967, nine months after its discovery, the only known pygmy possum developed paralysis and died (Green and Osborne 1994: 58). Three years later, several more individuals were found in the Snowy Mountains, about 100 km (62 miles) from the original site. Mountain pygmy possums are now known to occur in four isolated populations in the Snowy Mountains of northeastern Victoria and southern New South Wales (Groves 2005). They are found at elevations ranging from 1,373 m to close to the summit of Mount Kosciusko, at 2,228 m (about 4,500–7,300 ft). Optimal habitats range from rocky areas of periglacial-block streams to heaths and riparian habitats (Green and Osborne 1994). Pygmy possums mate in the spring, slightly later in the Snowy Mountains than in the Victorian Alps. A litter of four is born in November, with young leaving the pouch in December. Young are weaned at 8 to 9 weeks of age. Breeding seems to be related to availability of the principal food, the Bogong moths.

The classic pattern of hibernation exemplified by the mountain pygmy possum is taken to the extreme by one member of the order Rodentia, the arctic ground squirrel (*Spermophilus parryii*) (fig. 7.5A). Entry into the hibernation period begins in autumn, with stepped periods of torpor alternating with periodic rewarming bouts reaching euthermy (fig. 7.5B). During the season of hibernation, arctic ground squirrels undergo 12 bouts of torpor with a duration of one to three weeks each. Bouts are separated by regular episodes in which the body temperature returns to normal levels (about 37°C). Brian Barnes of the University of Alaska measured core body temperatures of hibernating *S. parryii* and found that the minimum body temperature fell to −2°C. During this time, the temperature of the soil surrounding the burrow was −10° to −15°C (14°F–5°F), meaning that, unlike most small mammals, arctic ground squirrels must continually thermoregulate during hibernation (Barnes 1989, 1996; Barnes, Omtzigt, and Daan 1993; Boyer and Barnes 1999). The little brown bat (*Myotis lucifugus*) has also been shown to be capable of regulating body temperatures below freezing during hibernation (Hurst and Wiebers 1967).

Bats of the family Vespertilionidae are well-known hibernators, and most species in the temperate zones spend the winter in caves and mines for this purpose. The duration of hibernation for bats differs widely among

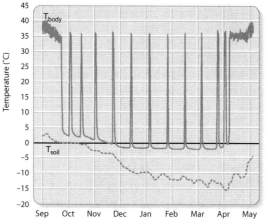

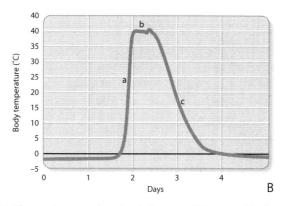

FIGURE 7.5. (A) The arctic ground squirrel (*Spermophilis parryii*) of northern Alaska undergoes seasonal hibernation. (B) In the upper graph, note the numerous bouts of torpor and arousal episodes and the extreme decrease in core body temperature (T_{body}) during deep torpor. Each arousal episode (lower graph) includes three phases: *a*, rewarming; *b*, euthermia; and *c*, cooling. (See the text for more details.) *Sources:* (A) Øivind Tøien; (B) Feldhamer et al. 2007, adapted from B. B. Boyer and B. M. Barnes (*BioScience* 49:713–724, 1999).

species and within a species, depending on the geographic area. The little brown bat is one of the most abundant and widely distributed **insectivorous** bats in North America. This is truly a "little" bat, weighing between 7 and 14 g (0.3–0.5 oz) with a wingspan of 22 to 27 cm (9–11 inches). In the summer, little brown bats are commonly observed foraging close to lakes and streams where they feed mainly on emerging aquatic insects. Each bat may consume more than one-half its body mass in insects in a single night. Throughout this time of year, the bats undergo short bouts of torpor while they roost during the day.

In early autumn, *M. lucifugus* may accumulate body fat equal to one-third of its body mass. Also during this time, a phenomenon called "swarming" may occur, when large numbers of little brown bats congregate around cave entrances. The purpose of this behavior is not fully understood. Some scientists believe that swarming may represent a mating event that facilitates gene flow between bats from otherwise isolated summer colonies, and/or it may serve to familiarize that year's young with possible locations to hibernate. By late autumn in northeastern North America, many little brown bats gather in the caves and abandoned mine shafts that serve as hibernacula. In Michigan, for instance, bats may hibernate for seven to nine months, in sites characterized by above-freezing ambient temperatures, near 5°C (41°F) and often typified by high humidity. A hibernating bat does not drink, except possibly during periodic arousals, so high ambient moisture is needed to prevent desiccation. Bats rarely hibernate where relative humidity is below 80% (Kurta 1995). Throughout the lengthy bouts of torpor during hibernation, the bats maintain a body temperature roughly 1°C above the ambient. During this time, metabolism also drops greatly, resulting in lowered respiration and heartbeat. For example, the heartbeat of a flying little brown bat may reach 1,368 beats/minute, but with a rectal temperature of 7°C during hibernation, its heartbeat decreases to only 20 beats/minute (Kallen 1977).

Bats exhibit the typical pattern of hibernation, involving a series of prolonged torpor bouts interrupted by brief arousals (French 1985; Thomas, Dorais, and Bergeron 1990). Bats spontaneously arouse every one to two weeks, achieving a normal body temperature that they then maintain for several hours. Arousals require a large quantity of energy—an amount equal to that required for 60 days of continuous hibernation. Don Thomas and coworkers studied energy budgets of little brown bats hibernating in an abandoned mine in Quebec, Ontario. They found that the metabolic rate required to defend a euthermic body temperature at typical hibernation temperatures can be up to 400 times that required during torpidity (Thomas, Dorais, and Bergeron 1990). For bats such as little brown bats occupying hibernacula in northeastern North America, each arousal requires costly energy production to increase body temperature as much as 35°C (95°C) and then defend the high body temperature for up to several hours. Hibernating bats achieve this increase in body temperature through a means of heat production called **nonshivering thermogenesis,** which does not involve muscle contraction. In most hibernators, the site of nonshivering thermogenesis is a specialized tissue known as **brown adipose tissue,** or brown fat. Brown fat is highly vascularized and well innervated and is capable of a very high rate of heat production. Deposits of brown fat are principally found in the interscapular regions (between the shoulder blades) in close proximity to blood vessels and vital organs (Hyvärinen 1994). During an arousal from hibernation, brown fat serves as a miniature internal "blanket" that overlies parts of the systemic vasculature and becomes an active metabolic heater applied directly to the bloodstream (fig. 7.6). Little brown bats have 13 separate deposits of brown fat strategically situated for effective and rapid transfer of heat to the blood (Rauch and Hayward 1969). Bats augment this nonshivering thermogenesis with shivering. I discuss the function of brown fat in further detail in Chapter 8.

Periods of arousal account for most of the energy used during hibernation—accounting for 70% or more of the total energy expenditure of a Richardson's ground squirrel (*Spermophilus richardsonii*), for example, during the hibernation season (Wang 1978, 1979). This is a seemingly wasteful expenditure of energy, and the function of the arousals is not well understood (Heller et al 1993). Some scientists contend that arousals are related to retention of memory and that this retention of memory requires slow-wave electrical activation that does not occur at the low body temperatures characteristic of hibernation. Therefore, periodic arousals are necessary to provide the conditions appropriate for slow-wave electrical activation and memory retention. Further, periodic arousals may permit hibernators to assess whether environmental conditions are conducive to emergence, or they may be required for conducting some endogenous repair functions that require mRNA and protein synthesis, which are not possible at a low body temperature (Heller and Ruby 2004). Due to the high energetic cost of arousal, there must be some evolutionary benefit for this phenomenon; as yet, however, it remains a mystery.

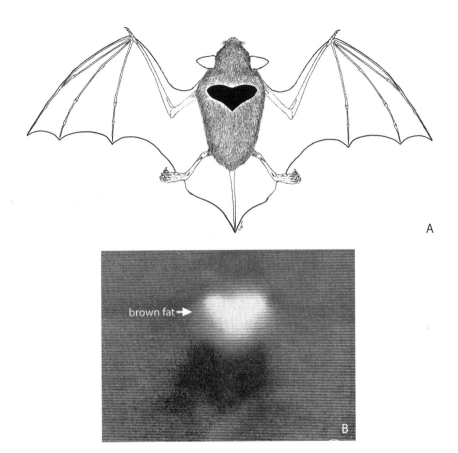

A

B

FIGURE 7.6.   (A) Position of interscapular brown fat in the big brown bat (*Eptesicus fuscus*). (B) Photograph taken with heat-sensitive film, showing the high temperature of interscapular brown fat in the big brown bat during arousal from hibernation. *Sources:* (A) Carie Nixon, Illinois Natural History Survey, adapted from J. S. Hayward and C. P. Lyman, *Mammalian Hibernation* (3d ed., Elsevier, 1967); (B) Hayward and Lyman 1967.

Hibernation is a season of limited activity and energy conservation, but it can also be a deadly season for the mammals that render themselves vulnerable in this way. The woodland jumping mouse (*Napaeozapus insignis*), the meadow jumping mouse (*Zapus hudsonius*), and the western jumping mouse (*Z. princeps*) (fig. 7.7A) are North American hibernators. The genus name *Napaeozapus* derives from three Latin roots and is liberally translated as "a woodland nymph with very large hind feet." Woodland jumping mice are most abundant in cool, moist hemlock-hardwood forests of mountain-

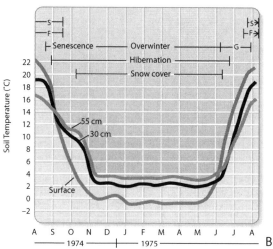

FIGURE 7.7. (A) Western jumping mouse (*Zapus princeps*) of the Wasatch Mountains of Utah. These mice undergo a period of hibernation from early September to late July. (B) Daily average soil temperatures (°C) at the surface and at 30 cm and 55 cm depth, at an elevation of 2,900 m in Albion Basin, Alta, Utah, from 1 August 1974 to 15 August 1975. The lines are plotted through the weekly mean. *G* = plant growth; *F* = flowering; *S* = plants setting seed. *Sources:* (A) Courtesy of The Roger W. Barbour Collection at Elon University; (B) Feldhamer et al. 2007, data from J. A. Cranford (*Journal of Mammalogy* 59:496–509, 1978).

ous regions of eastern North America, where they reside near streams that support a lush, low, woody vegetation. Shortening day lengths of autumn probably signal the mice to eat more to accumulate large fat reserves. In two weeks, the woodland jumping mouse may put on fat equaling one-third of its body mass. In Pennsylvania, *N. insignis* begins hibernation during October. The mice retreat (either singly or in pairs) to their underground hibernation nest composed of dry leaves and grass on a well-drained hillside. Here, the mouse buries its nose in its belly fur and draws up its hind legs along the face to form a tight ball, encircled by its long tail. A significant cause of death in woodland jumping mice is winter mortality. It seems that mice born in early autumn are likely to die during hibernation because they are unable to accumulate adequate fat reserve to endure the long winter— up to 75% of the mice that enter hibernation do not emerge in the spring. Other factors contributing to this grim statistic include predation, spring flooding, and severe cold stress.

In the Wasatch Mountains of Utah, the period of hibernation of *Z. princeps* ranges from early September to late July, depending on elevation of the hibernacula (Cranford 1978). A summer-active period of about 87 days was found to span the period between snowmelt in early summer and the beginning of the autumn snowfall season. Mice hibernated at an average depth of 59 cm (23 inches), with no food caches. The mean soil temperature during hibernation was 4.6°C (40.2°F), and emergence from hibernation was cued by increasing soil temperature (fig. 7.7B).

CHAPTER EIGHT

Coping with Cold

....

Winter is a potentially stressful time for northern mammals (and for mammals living at extreme southern latitudes), and they employ a wide array of adaptations to cope with the food shortages and cold stress that the season brings. The key to survival in the cold is to maintain energy balance (Wunder 1984). Two major features determine the energy needs of small mammals in the natural environment: the physical environment and the activities of the mammal (i.e., type and level of behavior, growth, or production of young) (Feldhamer et al. 2007: fig. 9.4). To meet their energy demands, small mammals must acquire energy—they must feed. Given a finite source of food during winter, mammals allocate energy, initially, to **thermoregulation**. Temperature regulation is essential for mammals to perform their normal functions. For example, a mammal may freeze to death even with abundant food if it cannot assimilate that energy and produce heat fast enough to balance heat loss. But even a mammal with a high heat-generating capacity will perish if there is not enough food in the environment to meet its needs. In sum, animals residing in cold regions must have sufficient food available and sufficient turnover capacity to meet their thermoregulatory needs. As mentioned in Chapter 6 (see table 6.1), mammals have evolved mechanisms of avoiding and resisting cold. Our discussion now focuses on the diverse tactics that mammals employ in surviving cold.

Insulatory Changes
....

To maintain body temperature within the **thermoneutral zone** in cold environments, a mammal (or bird) must reduce the rate of heat loss to its surroundings. The best way to reduce **thermal conductance** is to possess fat, fur, or feathers as an insulating layer between the body core and the environment. Insulation value increases with the thickness of fur and is at a maximum in mammals such as Dall sheep (*Ovis dalli*), wolves (*Canis lupus*), arctic foxes (*Vulpes lagopus*), and snowshoe hares (*Lepus americanus*). Mammals can also reduce thermal conductance in many other ways, besides insulation: by having a large body, changing peripheral blood flow, or modifying behavior, such as curling the body, **piloerection** (the fluffing of fur), nest building, and huddling in groups. Hair is a principal means of conserving energy.

Arctic foxes are champion survivors in the cold, coping with the vicissitudes of the harsh arctic environment by combining numerous **adaptations,** including lush **pelage.** These small canids, weighing between 3 and 4.3 kg (6.6–9.4 lb), have a **circumpolar** distribution that includes the arctic and tundra zones of the Holarctic region. *V. lagopus* is the smallest member of the order Carnivora that remains active during the long arctic winter. The pelage is twice as long in winter as in summer, and 70% of the winter pelage contains a very fine, insulative layer of underfur (Underwood and Reynolds 1980). Thick insulation confers the ability to extend the limit of the **lower critical temperature,** and the arctic fox holds the record in this category. This small canid has a lower critical temperature of −40°C (−40°F); at an environmental temperature of −70°C (−94°F), arctic foxes are known to elevate their **metabolic rate** only 50% above normal while maintaining **euthermy.** During bouts of severe cold, arctic foxes decrease activity, lower metabolic rate, and seek shelter in temporary dens or snow burrows to reduce thermal conductance (Frafjord 1992). In addition to its impressive insulating fur, the fox's other adaptations to cold include a compact body, thermoregulatory mechanisms in the feet, and the ability to reduce **metabolism** during cold conditions or food shortage (Scholander et al. 1950; Prestrud 1991; Merritt 1995; Merritt, Zegers, and Rose 2001; Audet, Robbins, and Larivière 2002). Foot temperature is modified by vasoconstriction of arterioles in the skin, coupled with the presence of peripheral subcutaneous fat deposits; freezing of the footpads is prevented by increased blood flow to a capillary network in the peripheral skin pads. Arctic foxes also conserve energy in locomotion on the surface of snow—a "savings" attributable to their relatively large feet, which yield a low foot load.

Snowshoe hares of the Kluane region of the Yukon conserve energy in cold environments by modifications of their pelage insulation. The average resting metabolism of hares was found to be 20% lower and thermal conductance 32% lower in winter than in autumn. **Guard hairs** were 36% longer and 148% denser in winter than in autumn (Sheriff et al. 2009).

There are some trade-offs associated with increases in insulation, especially for the smallest of mammals. The insulating value of fur is a function of its length and density; some small mammals, such as lemmings (*Dicrostonyx, Lemmus*), have very long fur relative to body mass. A thick coat on predators, such as weasels, however, would compromise their ability to forage in narrow crevices or **subnivean** spaces during winter months. Scholander (1955) contended that small mammals could not accumulate enough insulation in the form of fat or hair to cope with low winter temperatures without compromising agility. Research documenting the importance of pelage insulation in small mammals has focused primarily on mice and shrews (Churchfield 1990; Ivanter 1994). When these groups were faced with cold stress, heat loss was reduced by increasing pelage insulation by 11% to 19% during winter months. For small mammals, pelage changes may be useful in conserving energy and compensating for the large difference between body and environmental temperatures during winter. As we will see later in the chapter, small mammals that do not have a thick, long pelage take advantage of stable microclimatic conditions belowground and may employ social thermoregulation to cope with the cold. Larger mammals, however, benefit greatly from insulation in the form of fat and fur.

Countercurrent Heat Exchange: The Miraculous Net

The fur thickness of northern mammals may increase as much as 50% during winter months. Appendages such as legs, tail, ears, and nose, however, are potentially great heat "wasters." They cannot be well insulated, and to prevent heat loss from these sites, arctic land mammals permit appendage temperatures to decrease, often approaching freezing point. Appendages must be supplied with oxygen and kept from freezing by circulating blood. This is done, without constant loss of heat from the extremity, by a process called **countercurrent heat exchange**, a form of peripheral **heterothermy**. As with the concept of thermal windows described in chapter 9, countercurrent heat exchange is best exemplified with large mammals. This physiological mechanism shunts blood through a heat exchanger (the **rete mirabile**, or "miraculous net") that intercepts the heat on its way out and maintains the extremity at a considerably lower temperature than the core—the net result is a reduction of heat loss (fig. 8.1A). As warm arterial blood passes into a leg, for example, heat is shunted directly from the artery to the vein and then carried back to the core of the body. As a consequence, the appendages of many northern mammals are maintained at comparatively cold temperatures (fig. 8.1B). Anatomical studies have shown that the footpads of arctic canids have a massive arteriovenous plexus through which blood flow to and heat loss from the footpads are controlled (Henshaw, Underwood, and Casey 1972). To keep the extremities soft and flexible at such low temperatures, the fat in the feet of northern mammals must have a very low melting point, perhaps 30°C lower than that of other body fats (Storey and Storey 1988). Countercurrent heat exchange is well developed in the tail of beavers (*Castor canadensis*) and the ears of Japanese hares (Cutright and McKean 1979; Ninomiya 2000). In air at low temperature, loss of heat from the tail may be reduced to less than 2% of the resting metabolic rate. Then, when faced with higher temperatures, more than 25% of the heat produced at resting metabolism may be dissipated through the tail (Coles 1969).

The extent to which such vascular arrangements can work for animals coping with heat stress rather than cold stress depends on several properties of the countercurrent exchanger. Cooling of the outgoing arterial blood in the appendages is promoted by a high degree of contact between arteries and veins and by a relatively slow rate of blood flow through the exchanger. The arms of humans, the flippers of dolphins, and the limbs of tropical mammals such as sloths have a unique countercurrent exchange system. They possess two sets of veins: one superficial and distant from the major arteries, and the other deep and part of the exchange system. By adjusting the return of blood along these two venous systems, the mammal can emphasize heat dissipation or conservation in its extremity, according to its thermal needs. The large ears of black-tailed jackrabbits (*Lepus californicus*) are excellent heat dissipators (fig. 8.2B). Most excess heat generated during activity may be lost by dilation of the arteries in the ears (Hill, Christian, and Veghte 1980). The thermal conductivity of the heat-exchange system of jackrabbits is reported to be about 10 times greater for animals at 23°C (73.4°F) than at 5°C (41°F).

Reduced Level of Activity

Daily and seasonal temperature changes influence the activity patterns of winter-active small mammals. Nonhibernating species residing in seasonal

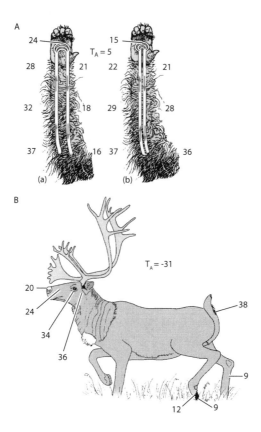

FIGURE 8.1. (A) Diagram representing circulation in the forelimb of a mammal, showing hypothetical temperature changes (°C) of the blood (a) in the absence and (b) in the presence of countercurrent heat exchange. Arrows indicate the direction of blood flow. In (b), the venous blood takes up heat (thus cooling the arterial blood) all along its path of return because, even as it becomes warmer, it steadily encounters arterial blood that is warmer yet. T_A = ambient temperature. (B) Regulation of body temperature in the caribou (°C). Temperature regulation is accomplished in part by countercurrent heat exchange. An intricate meshwork of veins and arteries acts to keep the temperature of the legs near that of the environment, so heat is not lost from the body. *Sources:* (A) adapted from Feldhamer et al. 2007, data from R. W. Hill and G. A. Wyse, 1989, *Animal Physiology* (2d ed., Harper and Row, 1989); (B) adapted from Feldhamer et al. 2007, data from J. F. Merritt, in *Arctic Life: Challenge to Survive* (ed. M. M. Jacobs and J. B. Ricardson III, The Board of Trustees, Carnegie Institute, 1983).

FIGURE 8.2. (A) The short ears of the snowshoe hare (*Lepus americanus*) assist in conserving heat in the harsh cold encountered in northern forests. (B) The large ears of the black-tailed jackrabbit (*Lepus californicus*) act as excellent dissipaters of heat when hares are faced with extreme temperatures encountered in desert environments. *Sources:* (A) Hal S. Korber; (B) Troy L. Best, ASM Image library.

environments must secure adequate nourishment during winter to maintain a constant, high body temperature. Among winter-active mammals, shrews are characterized by rapid heat loss due to their large surface-area-to-volume ratio, high metabolic rate, and resulting high caloric requirements. We would therefore predict that they are poor candidates for enduring cold stress. The northern short-tailed shrew (*Blarina brevicauda*) of eastern North America conserves energy by using **cached** food reserves and avoids cold environmental temperatures by restricting foraging and nesting to zones with stable microclimates. Because temperature changes influence the activity of invertebrates in the soil, foraging by soricid predators may also be affected by temperature. Churchfield (1982) used time-lapse photography to analyze the influence of temperature on the activity and food consumption of *Sorex araneus*. Shrews were active throughout the day and night, with peaks in activity every one to two hours. Activity outside the nest in summer was

28% of total activity, but this dropped to 19% during winter, concomitant with a decrease in food consumption. I. G. Martin (1983) showed an annual activity range for *B. brevicauda* of 7% to 31% of the day. Daily activity during winter was greatly reduced, averaging only 11.6% in the coldest months. *Blarina* spends 80% to 90% of the day resting in its nest at a low metabolic rate, sleeping for long periods and being intermittently highly active. A reduction in caloric intake and foraging activity during winter represents a survival tactic for coping with cold. In contrast, the seasonal metabolism of *B. brevicauda* increases from a low in summer to a maximum in autumn and winter. Northern short-tailed shrews may depart from the typical metabolic profile of shrews because of their proclivity for hoarding food.

Unlike the case for *Blarina*, food is severely limited for many rodents during winter months. For species of *Peromyscus* inhabiting northern latitudes, **torpor** combined with communal nesting is important to conserving energy. For example, in western Kansas, *P. leucopus* was found to spend 72% of the day in the nest during winter, compared with 28% of the day in summer. During winter, when the cost of thermoregulation is highest and food resources are lowest, the available energy must be used to maintain metabolism. Energy loss can be minimized in part by curtailing locomotor activity. For voles (*Microtus pennsylvanicus*), cold temperatures during winter, especially at night, may select for more **diurnal** activity. In winter, voles demonstrate increased movements on warm days or forage below a mantle of snow. When there is no insulating blanket of snow, low temperatures stimulate the use of nests and inhibit activity in voles. Hottentot golden moles (*Ambysomus hottentottus*) living at high elevations (1,500 m, or 5,900 ft) in the subtropical savannah of South Africa are exposed to cold conditions and shortages of food during winter. These **fossorial** mammals cope with seasonal stressors by restricting activity to shorter, more intense periods and decreasing thermal conductance by increasing pelage insulation (Scantlebury et al. 2005).

Small mammals such as shrews and voles, active during midwinter, do not undergo physiological heterothermy. The thermal regime of the foraging zone, therefore, is crucial in dictating energy budgets in these winter-active mammals. The microclimatic conditions of the foraging and nesting sites of shrews, voles, and mice have been examined during winter in Michigan, Ontario, and Pennsylvania. In mixed deciduous forests, many small mammals forage in tunnels within soil covered by a rich layer of leaves.

During winter, this foraging zone provides a stable, comparatively warm thermal environment. Snow covering the ground also provides additional insulation. Although ambient temperatures may reach −29°C (−20.2°F) in mid-January, the minimum temperature at the soil–leaf litter interface is about −4°C (24.8°F), and is as high as 1°C within a subsurface tunnel. Snow cover is an integral part of the life of small mammals. The presence of a sufficient depth of snow, called the heimal threshold, insulates the subnivean environment against widely fluctuating outside temperatures. High mortality rates among southern red-backed voles (*Myodes* [formerly *Clethrionomys*] *gapperi*) and white-footed mice (*P. leucopus*) during mid-winter are attributable to a lack of snow cover to insulate the forest floor. The period of autumn freeze is also crucial to the survival of small mammals, given the great fluctuations in temperatures in their foraging zone. While studying the population ecology of small mammals in the Front Range of the Rocky Mountains, I found a high mortality of voles and mice during late autumn (Merritt and Merritt 1978). At this time, snow thickness was insufficient to insulate soil against fluctuating ambient temperatures, which may reach −20°C (−4°F) in mid-November. Mortality of small mammals leveled off during winter months because of a blanket of snow that insulated their subnivean foraging zone against ambient temperatures. The period of snow cover typically lasts approximately eight months, and the snow cover reaches depths up to 2.5 m (about 8 ft) in many spruce-fir forests of the Rocky Mountains in May. As a result, small mammals are well insulated and thus not subjected to fluctuating ambient temperatures during midwinter.

Reduction of Body Mass (Dehnel's Phenomenon)

Small mammals, notably voles and shrews inhabiting seasonal environments, are reported to undergo a general decline in body mass during winter. This overwinter mass decline, called **Dehnel's phenomenon**, is thought to confer the adaptive advantage of decreasing caloric needs at a time when food resources are limited (Feldhamer et al. 2007). For shrews, the reduction of body mass and length is accompanied by a reduction of brain mass and skull depth, as well as a decline in size of most internal organs. All these changes may lead to a decrease in **mass-specific metabolic rate** and food consumption during winter. Some northern species—namely, collared lemmings and northern short-tailed shrews (*B. brevicauda*)—depart from this trend and gain mass during winter. *Peromyscus* does not conserve energy in this way during winter. Its ability to undergo torpor coupled with communal nesting, food hoarding, and use of elaborate nests all aid considerably in conserving energy during winter. The cues for such fluctuations in mass are complex and include photoperiod (length of daylight), environmental temperature, and availability of food. The common shrew (*S. araneus*) in Eurasia also demonstrates body mass decreases during winter. Declines in body mass can be significant—*S. araneus* in Great Britain, Poland, and Finland was found to lose up to 45% of body mass, and *S. cinereus* in North America showed a decline of 53% in body mass from early summer to winter (Merritt 1995). As noted above, small size during winter confers an energy advantage, presumably as a means of reducing food requirements (Lovegrove 2005). A small mammal eats less food and has a greater assimilation efficiency than a large one; consequently, it can reduce foraging time in cold conditions and thus conserve energy. Although a small body mass in winter decreases food requirements, it also reduces cold tolerance due to an increase in surface-area-to-volume ratio.

The survival mechanisms discussed so far in this chapter have one thing in common: they pertain to animals that must forage for food to maintain a high body temperature year-round. However, many small mammals maintain a very low body temperature to cope with cold and heat and may not consume food for six to nine months of the year!

Fat Tails

Tails of small mammals vary greatly in structure and function. Kangaroo rats, pocket mice, gerbils, jerboas, and long-tailed dunnarts sport long tails that act as counterbalances as these small mammals move along the desert floor. **Arboreal** species such as honey possums and pygmy possums of Australia, mouse opossums of South America, and golden mice of North America have **prehensile** tails as an adaptation for arboreal life. In addition to these locomotor functions, the tails of some small mammals have a role in meeting energy requirements in the form of energy storage. It is well known that marine mammals such as whales, walruses, and seals possess a particularly effective insulator—namely, blubber. Consisting of collagen and elastic fibers embedded in an incompressible matrix of adipocytes (fat cells), blubber is located below the dermis of the skin. For marine mammals, in addition to providing insulation, blubber acts to streamline the body, adjust buoyancy, provide an energy storage site, support thermoregulation, and, possibly, provide a hydrodynamic function. Small mammals do not have blubber. As mentioned earlier, they reduce thermal conductance by increasing pelage length and density and putting on minimal amounts of subcutaneous white adipose tissue (white fat). (Soricids put on very little white adipose tissue but rather rely on **brown adipose tissue** for nonshivering thermogenesis, as discussed later in the chapter.) White adipose tissue is typically considered an organ associated with the storage of lipid that acts to insulate mammals inhabiting cold environments. Recent research, however, indicates that white adipose tissue also serves as an important endocrine organ associated with metabolic and physiological regulation (Trayhurn, Beattie, and Rayner 2000). White fat secretes a signal protein called leptin that is important in processes of energy balance, reproduction, and immunity (Zhang et al. 1994).

Several species of small mammals found in variable environments and exposed to cold stress have **incrassated** tails—they store fat in the base of their tail to support periods of torpor (fig. 8.3). This adaptation is found in species inhabiting a wide range of ecosystems throughout the world. Representatives include star-nosed moles of North America; Chilean shrew opossums, monito del montes, Patagonian opossums, and fat-tailed opossums of South America; white-toothed shrews of Europe and Africa; and fat-tailed pseudantechinus of Australia—to mention just a few. Among the "insectivores," the star-nosed mole (*C. cristata*) has a tail that is almost as long as its body, which appears to store fat that will be utilized during the breeding season when energy demands are unusually high. In addition, species of white-toothed shrews (*Crocidura*) that inhabit arid regions of Africa have very fat tails, which may aid in coping with the dry season when food is in short supply.

Among the small mammals inhabiting South America, incrassated tails are common in many members of the subfamily Didelphinae: the mouse opossums, fat-tailed opossums, and Patagonian opossums. Mouse opossums occur in arboreal and **terrestrial** environments from Mexico through South America, inhabiting many different communities from sea level to the mountains. They are absent only from the high Andes Mountains, the Chilean desert, and Patagonia. In Patagonia, mouse opossums are replaced by Patagonian opossums, which have the most southerly distribution of any didelphid **marsupial**. The tail in many of these small opossums becomes thickened

FIGURE 8.3. Southern mouse opossum (*Thylamys pusilla*) of South America, with in-crassated tail. An incrassated tail, or "fat tail," is found in many species of small mammals in seasonal environments. Such tails are thought to be adaptive for sur-vival, as a site for storing fat to support periods of torpidity. *Source:* Rexford D. Lord.

near the base due to a seasonal accumulation of fat. For example, the base of the tail of the elegant fat-tailed mouse opossum (*Thylamys elegans*) can measure up to 1 cm (0.4 inches) in diameter during incrassation. As in other southern species of opossums, the tail of this opossum fattens in winter, reaching its maximum thickness in August when the animal enters torpor.

Shrew opossums (family Caenolestidae) inhabit the Andean region of western South America, from southern Venezuela to southern Chile. Of the six species of caenolestids, only the long-nosed caenolestids (*Rhyncholestes raphanurus*) have fatty protuberances at the base of their tail (Tyndale-Biscoe 2005). *R. raphanurus* inhabits forests of southern Chile and adjacent Argentina, including Chiloe Island. Individuals captured in late summer and autumn show an enlarged area of stored fat at the base of their tail as an indication of preparation for winter torpor or **hibernation**. We know very little of the biology of these elusive and secretive shrewlike marsupials.

Once thought to belong to the same family as the American opossums, the diminutive monito del montes, or "little monkey of the mountains" (*Dromiciops gliroides*), is now considered the only living member of an otherwise extinct order, the Microbiotheria. The two extant species of monito del montes, or colocolos (family Microbiotheriidae), weigh less than 30 g (1 oz). Both have short fat tails and are known to hibernate during winter months. They inhabit the humid forests of beech (*Nothofagus*) and bamboo (*Chusquea*) of south-central Chile and adjacent Argentina. Monitos are **nocturnal** and arboreal and make distinctive, round nests in fallen logs, tree cavities, and thickets. Nests are often lined with leaves of water-resistant Chilean bamboo. Microbiotheriids forage primarily for insects, especially beetle larvae and pupae, but occasionally may consume herbaceous material. Prior to hibernation, fat accumulates in the base of the incrassated, prehensile tail. Enough fat may be stored in a week to double an individual's body mass.

Social Thermoregulation

Many mammals reduce the differential between body surface and ambient air temperatures by constructing lodges or burrows that provide comparatively stable environmental temperatures (Scholander et al. 1950; A. B. Stephenson 1969; Buech, Rugg, and Miller 1989). In northern regions, nonhibernating small mammals, particularly those of the rodent families Cricetidae and Sciuridae, construct elaborate nests and engage in communal nesting (Feldhamer et al. 2007). Dale Madison (1984) outlined four distinct advantages for group nesting The first advantage is avoiding predators and improving defense—"safety in numbers." Another often cited benefit pertains to the efficiency of food acquisition; this advantage is clearly seen in the "social carnivory" employed by wolves when hunting prey. The third factor presented to account for the evolution of social grouping is scarcity of habitats. This factor seems more speculative, suggesting that small mammals may form aggregations in landscapes that provide "safe ground" away from flooding and other habitat perturbations. The fourth and most important, and certainly the best studied, advantage to group nesting pertains to energy conservation. The greatest gain from huddling should accrue for small mammals with a large surface-area-to-mass ratio and a limited capacity for increasing the insulation value of their pelage. Not only small mammals use huddling, however: 23 raccoons (*Procyon lotor*) were found occupying one winter den. During winter, female striped skunks (*Mephitis mephitis*) form a communal nest with other females or with a single male. Communal denning has adaptive value for winter survival and reproductive success, especially in northern latitudes. For example, in Alberta, an average occupancy rate of 6.7 striped skunks per communal den was found, with females more common in dens than males (Gunson and Bjorge 1979).

Both nesting and huddling conserve body heat through reductions in thermal conductance. Huddling in groups reduces each individual's exposed surface, thus reducing cold stress and the metabolic requirement for heat production. Studies traditionally employed laboratory-based calorimetry to demonstrate the energy savings of communal nesting during winter. More recent studies have linked laboratory experimentation with field-derived data on nesting habits by using radiotracers, radiotelemetry methods, and oxygen-consumption techniques. When muskrats huddle in a group, a major part of each individual's body surface is in contact with a neighboring animal (Bazin and MacArthur 1992) (fig. 8.4). Curling and retracting the extremities reduces heat loss, and microclimatic modification in the form of an elaborate, well-insulated nest belowground and under snow cover (subnivean) adds greatly to energy savings. Ronald Bazin and Robert MacArthur (1992) studied the energy savings accrued by muskrats huddling in a shelter during winter near Winnipeg, Manitoba. For muskrats exposed to temperatures of −10°C and 0°C (the lowest winter microclimatic temperatures occurring in muskrat shelters), Bazin and MacArthur recorded the metabolic rate of an aggregation of four muskrats as averaging 11% to 14% below that of a single individual in a shelter. In earlier studies using radiotelemetry on muskrats inhabiting marshes near Manitoba, MacArthur and Mike Aleksiuk (1979) found that up to six muskrats may inhabit the same winter lodge; the resting metabolic rate of a group of four muskrats huddling in such a lodge during winter, with an environmental temperature of −10°C (14°F), demonstrated up to a 13% energy savings over that of a single animal. In northern latitudes, aggregations of at least six muskrats are common.

The classic work by John Sealander (1952) showed that at low temperatures, *Peromyscus* individuals formed a "communal" group. Those at the bottom of the group enjoyed temperatures well above ambient levels, and by continually shifting position, each mouse in the huddle was periodically rewarmed and thus avoided **hypothermia**. Because heat loss by conduction or convection varies in direct proportion to the amount of surface area exposed, the energy savings of huddling can be easily calculated. Vickery and Millar (1984) provided a model for predicting the energy advantages and disadvantages of huddling behavior. When applying their model to data on *Peromyscus*, they found that huddling confers a distinct energy savings for

FIGURE 8.4. Many small mammals, such as these muskrats (*Ondatra zibethicus*), conserve energy in cold conditions by employing communal nesting. (A) Close and (B) loose aggregation responses of muskrats exposed to a temperature of 5°C (41°F). *Source:* Feldhamer et al. 2007, courtesy of Robert MacArthur.

mice subjected to ambient and nest temperatures well below the thermoneutral zone.

Rodents construct elaborate nests of grasses and herbs under the litter, on the ground, or within a hollow tree or log. During the winter in northern regions, these nests are commonly located in the subnivean environment, which aids in insulating the nest site from fluctuating supranivean (above-snow) temperatures. The physical structure and communal use of beaver lodges in southeastern Manitoba were assessed by Dyck and Mac-Arthur (1993). Winter air temperatures outside the lodges reached a low of −41.4°C (−42.5°F), but temperatures within the chambers of occupied lodges did not fall below 0°C. The mean monthly temperature of the nesting chamber consistently exceeded the mean monthly exterior air and water temperatures. The ameliorated microclimate within lodges also facilitates periodic rewarming of foraging beavers, thus minimizing thermoregulatory costs during rest.

Jerry Wolff and Bill Lidicker, working in central Alaska, found that taiga voles (*Microtus xanthognathus*) display some very interesting social behaviors adaptive for overwinter survival (fig. 8.5). During winter months, the voles lived in underground nests composed of communal groups of 5 to 10 individuals. The communal nests under the snow remained between 7°C (44.6°F) and 12°C (53.6°F) warmer than ground temperature and as much as 25°C (77°F) warmer than ambient temperatures. Furthermore, nests were not completely vacated at any time, so that foraging voles returned to a warm nest (Wolff 1980; Wolff and Lidicker 1981). Also, during winter months, voles utilized underground food caches of horsetail (*Equisetum arvense*) and fireweed (*Epilobium angustifolium*) rhizomes. These **middens** were rather large, measuring 1 m² (10.7 ft²) and 20 to 30 cm ((7.8–11.8 inches) high and containing up to 3.6 kg (about 1 bushel) of rhizomes—representing 90% of winter food for the voles.

Huddling by deer mice and voles may reduce energy requirements in the cold by as much as 16% to 36% (Vogt and Lynch 1982; Andrews and Belknap 1986). For *P. leucopus,* a combination of torpor and huddling of three individuals in a nest at an ambient temperature of 13°C (55.4°F) resulted in a daily energy savings of 74% compared with nontorpid, individual mice without a nest. Aggregate nesting during winter is common for species of *Peromyscus.* The average daily metabolic rate of *P. maniculatus* is also lower in winter. When *P. maniculatus* and *P. leucopus* are syntopic (live in the same locality) in the Appalachian Mountains of Virginia, radiotelemetry studies show that *P. maniculatus* prefers nesting high in large, hollow trees year-round, whereas *P. leucopus* uses both ground and tree nest sites in summer but shifts to underground nest sites in winter (Wolff and Durr 1986). Radiotelemetry studies have also demonstrated that the two species will coinhabit one nest during winter months.

Conservation of heat by a group of huddling animals is greatest when

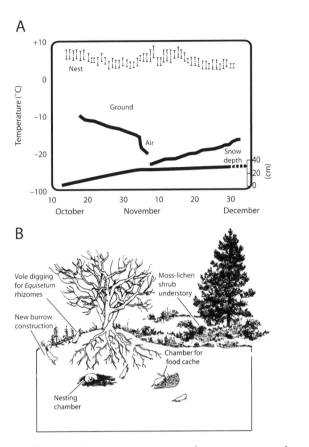

FIGURE 8.5. (A) Daily mean air temperature, ground temperature, and nest temperature (°C) for the taiga vole from 15 October to 1 December 1977, at a site 155 km (96 miles) northwest of Fairbanks, Alaska. The daily ranges of nest temperature are shown by the vertical bars (top). Increasing ground temperatures after 20 October were due to progressively deeper snow cover. (B) The communal winter midden-tunnel system and activities of a group of taiga voles. *Sources:* (A) Feldhamer et al. 2007; (B) data from J. O. Wolff and W. Z. Lidicker Jr. (*Behavioral Ecology and Sociobiology* 9:237–240, 1981).

the nest is well insulated. Researchers have evaluated the thermal capacity of nests by calculating their shape, composition, and wall thickness. The resistance of nests to heat loss can also be measured quantitatively (Wrabetz 1980). Interspecific differences in nest-building behavior were correlated with available microhabitats and nest site preferences (Layne 1969). J. A. King and colleagues demonstrated a geographic correlation in the amount of nest material used by *Peromyscus* (J. A. King, Maas, and Weisman 1964). Northern forms used more nesting material, under constant temperatures, than did southern forms. Sealander (1952) showed a seasonal difference in nest-building behavior: *P. leucopus* and *P. maniculatus* constructed more elaborate nests in winter than in summer, with winter nests conferring greater resistance to low temperatures. Using outdoor enclosures, researchers demonstrated that *P. leucopus* and *P. maniculatus* in northern New York State constructed larger nests during winter than at other times of the year.

Marmots are clearly challenged by the vicissitudes inherent in long, harsh winters. All 14 species reside in the northern hemisphere, mainly in montane regions where food is unavailable during the winter—and winters may last for seven months. Marmots are the largest true hibernators, ranging in size from 3.4 kg (7.5 lb) for the yellow-bellied marmot (*Marmota flaviventris*) of the Rocky Mountains to 7.1 kg (15.7 lb) for the Olympic marmot (*M. olympus*) of Washington State. These mammals are the most social hibernators of the squirrel family; most species of marmots hibernate in groups in a common burrow (a hibernaculum). Only the woodchuck (*M. monax*) leads a solitary existence (Armitage 1999). Walter Arnold (1988) studied the biology of a population of alpine marmots (*M. marmota*) in the Berchtesgaden Alps of Germany for seven years. The social groups consisted of a dominant territorial pair and up to 18 subordinate marmots of various ages. Arnold found that during the hibernation period, alpine marmots cuddle up to the young in communal burrows, preventing heat loss that would otherwise burn up fat reserves and cause weight loss and thus reduce the chances of survival for overwintering individuals. In autumn, marmots formed communal groups of sometimes five individuals huddling in a hibernaculum 1 to 3 m (3.3–9.9 ft) belowground. Communal groups were composed of the adult pair and several yearlings. They huddled tightly together in the nest and synchronized their regular changes between torpor and **euthermy**. The benefits of such huddles for marmots include a decrease in loss of body mass in winter and a significant reduction in winter mortality.

Most tree squirrels are solitary and euthermic during winter. Southern flying squirrels (*Glaucomys volans*), however, form huddles of up to 20 individuals (but groups of fewer than 10 are more common) in hollow trees to conserve energy. *G. volans* and *G. sabrinus* undergo periodic bouts of torpor during winter, due to extended periods of food shortage and low temperature (Muul 1968). A group of six southern flying squirrels in New Hampshire huddling in a wooden nest box and surrounded by temperatures of 6°C (42.8°F) reduced their energy expenditure by 36% (Stapp, Pekins, and Mautz 1991). This behavioral strategy, augmented by a nearby supply of hoarded nuts, enhanced survival during winter (Merritt, Zegers, and Rose 2001).

Increased Heat Production

As we have seen, mammals employ many different tactics of energy conservation to avoid cold stress. If cold stress persists, however, mammals must resist it by processes that generate heat and thus require energy. The major ways in which **endotherms** increase heat production are muscular activity and exercise, involuntary muscle contractions (shivering), and **nonshivering thermogenesis**. The most conspicuous mechanism is muscular activity from locomotion or shivering; the means need not be apparent to the observer, however, for it to increase the rate of heat production. Shivering is well documented for marsupials, humans, canids, felids, selected rodents, lagomorphs, and ungulates, as well as many species of birds. Small mammals residing in variable environments, however, employ nonshivering thermogenesis, which does not involve muscle contraction (Feldhamer et al. 2007). Brown adipose tissue (brown fat), the site of nonshivering thermogenesis, was first observed by Conrad Gesner in 1551 in the interscapular area of the Old World marmot (*Marmota alpina*). Brown fat is the primary thermogenic tissue of cold-adapted small mammals and is found in all hibernating mammals as well as in many small mammals that remain active throughout the winter. It is also well developed in many newborn mammals, including humans. Brown fat has been reported in many mammalian species, spanning nine orders: Dasyuromorphia, Chiroptera, Afrosoricida, Soricomorpha, Rodentia, Lagomorpha, Artiodactyla, Carnivora, and Primates (Feldhamer et al. 2007). Brown adipose tissue is found in eutherian (**placental**) mammals and probably evolved early in the adaptive radiation of this group. Monotremes do not have brown adipose tissue; however, researchers have found brown adipose tissue in some species of marsupials.

Unlike white adipose tissue, with cells characterized by a single large droplet of fat and a peripheral nucleus, brown fat cells contain many small droplets (multilocular) with a centrally located nucleus. In contrast to white adipose tissue, brown fat is highly vascular and well innervated. The cells contain many mitochondria, whereas white fat cells have comparatively few. Brown fat is capable of a far higher rate of oxygen consumption and heat production than white fat. The reddish brown color of brown fat is derived from iron-containing cytochrome pigments in the mitochondria, the essential components of the oxidizing-enzyme apparatus of brown adipose tissue. White fat serves primarily as insulation and as a storage site for food and energy. Brown fat, with its rich supply of mitochondria, serves as a miniature internal "blanket" that overlies parts of the systemic vasculature and becomes an active metabolic heater applied directly to the bloodstream. Deposits of brown fat can be rather diffuse but are principally found in the interscapular, cervical, axillary, and inguinal regions in close proximity to blood vessels and vital organs (see fig. 7.6).

Temperature receptors in the skin sense cold and send impulses to the preoptic area of the hypothalamus—the "mammalian thermostat" located in the brain. Impulses are then relayed along sympathetic nerves to the brown adipose tissue, where nerve endings release the neurohormone **norepinephrine**. In the brown adipose tissue, norepinephrine activates an enzyme (lipase) that splits triglyceride molecules into glycerol and free fatty acids. In the brown fat cell, mitochondrial respiration is "uncoupled" from the mechanism of adenosine triphosphate (ATP) synthesis so that the energy of oxidation of the fatty acids is dissipated as heat instead of being used for ATP synthesis. A special protein called **thermogenin** is responsible for the uncoupling (Himms-Hagen 1985). As shown earlier, when bats arouse from hibernation, the brown fat pad is much warmer than the rest of the body. The close proximity of Selzer's vein just beneath the interscapular brown fat permits rapid passage of warmed venous blood directly to the heart and brain, with a minimum of heat loss.

Research on northern small mammals shows that dramatic increases in metabolic rate in cold conditions are due to nonshivering thermogenesis. Nonshivering heat production usually tracks environmental temperatures, falling to the lowest rates in spring and summer, increasing in autumn, and peaking in winter. Typically, small mammals demonstrate a significant inverse relationship between nonshivering heat production and environmental temperature. Bruce Wunder and colleagues showed a 29% increase in oxygen consumption for prairie voles (*Microtus ochrogaster*) in winter compared with summer (Wunder, Dobkin, and Gettinger 1977). Northern red-backed voles (*Myodes* [formerly *Clethrionomys*] *rutilus*) exhibited a 96% increase in metabolism in winter. Species of *Peromyscus* commonly employ many different adjustments for survival, and they are adept at increasing metabolism during winter. Maximum metabolism for *Peromyscus* in Iowa and Michigan ranged from a 24% to 70% increase over summer rates. Shrews typically show very high rates of nonshivering heat production (J. R. E. Taylor 1998). For example, I found that for masked shrews (*Sorex cinereus*) in Pennsylvania, nonshivering thermogenesis in winter was almost twice that measured in summer—an increased capacity of 182% (Merritt 1995). Maximum metabolism, resting metabolic rate, and nonshivering thermogenesis were compared for 12 species of shrews of tropical and temperate regions (Sparti 1992). Maximum metabolism, and consequently improved cold tolerance, was pronounced in temperate species. There is a direct correlation between mass of brown fat and nonshivering thermogenesis. Dave Zegers and I conducted research in Pennsylvania (at Powdermill Biological Station) and found that monthly changes in the mass of brown fat are inversely related to minimum environmental temperature for many species of voles, mice, and shrews (Merritt and Zegers 2002).

CHAPTER NINE

Coping with Heat and Aridity

To survive in deserts, small mammals must cope with a variety of demanding environmental challenges, such as intense heat during the day, cold nights, paucity of water and cover, and a highly variable food supply. Desert ecosystems are widespread and abundant—35% of the earth is covered with deserts. Small mammals have successfully colonized desert ecosystems, as evidenced, for example, by the rich and diverse fauna of heteromyid and dipodid rodents in the deserts of North America and the Old World, respectively. Degen (1997) provides an excellent review of studies examining adaptations of mammals to desert environments.

As mentioned earlier, the challenges faced by small mammals in desert environments are even more severe than those encountered in cold regions; water as well as food is scarce, and the problem of temperature regulation is reversed. In considering cold stress in small mammals (in Chapter 8), I assessed the role of thermal radiation, conduction, convection, and other environmental influences on heat transfer between an individual and its surroundings, and then focused on mechanisms for reducing heat flow to the environment. In desert ecosystems, the gradient between internal and environmental temperatures is reversed. In some deserts, mammals may have to cope with air temperatures that reach 55°C (131°F) and ground temperatures exceeding 70°C (158°F). Unlike arctic small mammals concerned with conserving heat, desert mammals must dissipate or avoid heat to maintain euthermy. The discussion in this chapter focuses on the complex anatomical, physiological, and behavioral adaptations that enhance the survival of small mammals in desert ecosystems. Because water is essential for survival (it constitutes 70% of the body mass of mammals), the topic of water conservation in small mammals is a logical starting point.

For mammals inhabiting desert environments, the ability to concentrate urine is paramount and closely tied to kidney function. Because of the urine-concentrating ability of kidneys, mammals can produce urine that is hyperosmotic to (i.e., has a greater solute concentration than) blood plasma—up to 25 times the concentration of plasma. Not surprisingly, the highest urine concentrations are found in mammals of desert habitats, as well as in some species of bats (e.g., *Myotis vivesi*) subsisting on marine fish and crustaceans. The concentrating properties of kidneys in different species are closely related to the lengths of the loops of Henle and collecting ducts that traverse the renal medulla. The prominence of the medulla is commonly expressed as the relative medullary thickness (RMT; the thickness of the medulla relative to the size of the kidney), an important index of kidney **adaptation** (Sperber 1944). RMTs of mammals inhabiting arid areas are greater than those of mammals in more **mesic** environments (i.e., with more moisture). The relationship between RMT and maximum urine-concentrating capacity was first quantified by Schmidt-Nielsen and O'Dell (1961) and has proved useful as a means of comparing kidney function in mammals. The anatomy of the papilla of the medulla can be compared visually among species. In desert-adapted small mammals, the papilla may extend beyond the margins of the renal capsule into the ureter. This extension is pronounced in small desert rodents and dasyurid **marsupials**, shrews, golden moles, and bats, suggesting that these species have powerfully functioning kidneys. In contrast, aquatic mammals such as beavers, aquatic moles, water rats, and muskrats have very short loops (shallow papillae) and produce a less concentrated urine.

RMT can vary among closely related species residing in different habitats. Urine concentration in **insectivorous** bats also shows a relationship to habitat. The concentrating ability of the kidneys of North American bats, as indicated by the RMT, is significantly correlated with total annual precipitation in their habitats; however, difference in habitat aridity does not explain all of the variation among species, so other factors are also important. Research on dasyurid marsupials seems to be more conclusive. For example, studies of kidney structure of some small insectivorous and **carnivorous** dasyurids show that RMT varies greatly, ranging between 3.7 and

11.5, with species in arid habitats having higher values than those in semi-arid, mesic, and tropical habitats. Arid-adapted species such as Wongai ningaui (*Ningaui ridei*) and fat-tailed false antechinus (*Pseudantechinus macdonnellensis*) have the highest RMTs, while the alpine-dwelling dusky antechinus (*Antechinus swainsonii*) has the lowest.

Comparative studies of renal function and morphology in mammals indicate a direct relationship between the ecological distribution of a species and its ability to conserve urinary water. As noted above, the ability to concentrate urine in mammals is associated with long loops of Henle and tubules, which enhance the kidney's countercurrent exchange function. Species living in an arid habitat that tend to have kidneys better adapted for water conservation include members of the orders Dasyuromorphia (family Dasyuridae), Diprodontia (Macropodidae), Chiroptera, Cingulata (Dasypodidae), Rodentia (Muridae, Cricetidae, Octodontidae, Heteromyidae, and Sciuridae), and Lagomorpha (Leporidae). Most mammals lose water by excretion in the urine and elimination in feces, but these groups can produce concentrated urine and relatively dry feces. For example, the average values for maximum urine concentration in desert heteromyids are superior to those of most mammals and comparable to those of other desert-adapted small mammals, such as dipodids of Asia and North Africa and murids and dasyurid marsupials of Australia and southern Africa. The laboratory white rat can produce urine with twice the osmotic concentration that humans can achieve. Urine concentrations in insectivorous bats, as noted above, also show a strong relationship to habitat.

As expected, the amount of water loss in feces is low for desert mammals. The feces of Merriam's kangaroo rat (*Dipodomys merriami*) are more than 2.5 times drier than those of white rats (water content 834 and 2,246 mg/g of dry feces, respectively). Furthermore, heteromyids commonly decrease fecal water loss by assimilating more than 90% of the food they ingest.

Desert rodents (e.g., kangaroo rats, sand rats, and jerboas) produce urine concentrations of 3,000 to 6,000 mOsm/L (milliosmoles per liter); Australian hopping mice (*Notomys alexis* and *N. cervinus*) can produce urine concentrations exceeding 9,000 mOsm/L. Intuitively, we think of heteromyid rodents such as kangaroo rats and pocket mice as leaders in water conservation. Many other small mammals, however, such as pallid bats (*Antrozous pallidus*), fish-eating bats (*M. vivesi*), canyon and house mice (*Peromyscus crinitus* and *Mus musculus*), and golden hamsters (*Mesocricetus auratus*), have evaporative water losses equivalent to or greater than that of desert heteromyids. The spectacled hare-wallaby (*Lagorchestes conspicillatus*) inhabiting **xeric** grassy plains of northwestern Australia has the lowest mass-specific rate of water turnover reported for any mammal (S. D. Bradshaw, Morris, and Bradshaw 2001).

An additional way in which desert rodents minimize water loss is by producing a highly concentrated milk. The milk of *D. merriami* averages 50.4% water—a concentration comparable to that produced by seals and whales. Furthermore, in desert rodents, canids, and kangaroos, mothers reclaim water by consuming the dilute urine and feces of their young. This behavior may regain about one-third of the water originally secreted as milk.

Dietary Water Intake

Because free drinking water is not available for desert mammals, they must obtain water from other sources, such as from succulent plants or the body fluids of prey or from the dry food they consume. This requires subsisting on **metabolic water**, which is created in cells by the oxidation of food, especially carbohydrates. Desert mammals that consume succulent plants as a source of water include desert woodrats (*Neotoma lepida*) and cactus mice (*Peromyscus eremicus*) of southwestern North America, which eat large quantities of cactus (*Opuntia*) as a source of both food and water. Cactus is also a staple in the diet of other xeric-adapted mammals, such as northern pocket gophers (*Thomomys talpoides*) of the dry short-grass prairies of Colorado. An evolutionary development in gophers and cricetids such as woodrats is the ability to metabolize oxalic acid, an abundant compound in cactus that is toxic to other mammals. Instead of ingesting moist food such as cactus, other desert mammals rely primarily on dry seeds or halophytic plants (those that grow in salty soils), and their intake of water is minimal. Kangaroo rats, pocket mice of southwestern North America, and fat sand rats (*Psammomys obesus*) of the Saharo-Arabian deserts can subsist on dry food only and do not require water.

Halophytic plants of the family Chenopodiaceae form a staple in the diet of many desert-dwelling small mammals but are extremely high in salt concentration. Many small mammals that consume halophytes have kidneys that produce highly concentrated urine. Fat sand rats obtain water from the leaves of the saltbush (*Atriplex halimus*). These gerbillid rodents are unusual in being **diurnal** and wholly **herbivorous**, whereas other members of the family are **nocturnal** and **granivorous**. *Psammomys* scrapes off the outer surface of a leaf with its teeth before consuming it; negligible amounts of leaf are scraped from moist plants and substantial amounts from dry plants. The leaves contain up to 90% water but have high concentrations of salts and oxalic acid. To consume this plant material, the rat must produce urine with extremely high concentrations of salt and must be able to metabolize large concentrations of oxalic acid.

Jim Kenagy (1972) conducted a series of fascinating experiments on chisel-toothed kangaroo rats (*Dipodomys microps*) of the Great Basin shrub habitat in the Owens Valley of eastern California. The findings make an intriguing story. Chisel-toothed kangaroo rats (*D. microps*) and Merriam's kangaroo rats (*D. merriami*) coinhabit shrub habitats of western North America. *D. merriami* gleans seeds from the soil in typical kangaroo rat fashion, whereas *D. microps* climbs onto shrubs and harvests leaves of the chenopod (*A. confertifolia*), which it stuffs into its fur-lined cheek pouches and carries back to its burrow to be eaten or **cached**. Interestingly, *D. microps* will harvest leaves of *Atriplex* throughout the year, to the total exclusion of 11 other shrubs available on the site. The epidermis of *Atriplex* leaves is high in electrolyte concentration—sodium content is greater than potassium, which is the reverse of the situation in most plants; however, the more internal parenchyma of these leaves is low in electrolytes and high in starch. *D. microps* is able to consume the inner tissue by shaving off the peripheral epidermis, thus minimizing its consumption of salt. This kangaroo rat can perform such a task, which other kangaroo rats cannot, because it has lower **incisors** that are broad, flattened anteriorly, and chisel-shaped, permitting access to the plant's inner tissue. Because other, sympatric kangaroo rats (such as Merriam's) lack this adaptation, they cannot exploit this unique food resource and must rely on unpredictable seed crops. I recall reading about this work as a graduate student—what a unique specialization! This research by Jim Kenagy was one of the studies that initiated my interest in the physiological ecology of small mammals.

Desert carnivores and "insectivores" meet their moisture requirements by relying on their food rather than on free water. Southern grasshopper mice (*Onychomys torridus*) of hot, arid valleys and shrub deserts of southwestern North America consume primarily arthropods, including scorpions, beetles, and grasshoppers. Studies of water balance demonstrate that grasshopper mice can be maintained in the laboratory for more than three months on a diet of only fresh mouse carcasses. Grasshopper mice can survive the arid conditions of the desert because of their preference for animal foods high in water content. The mulgara (*Dasycercus cristicauda*) of arid regions of Australia eats primarily large insects, spiders, and scorpions; it is well adapted to life in the desert. Mulgaras' ability to produce urine with a high concentration of urea permits them to excrete large amounts of nitrogenous waste derived from their insectivorous diet in a relative small volume of urine, and thus they can survive entirely on metabolic water derived from food.

Kit foxes, badgers (*Taxidea taxus*), coyotes (*Canis latrans*), desert hedgehogs (*Hemiechinus auritus*), and fennecs (*Vulpes zerda*), like the mulgaras (*D. cristicauda*), can subsist on a meat diet with minimal supplementation by free drinking water. Fennecs are the smallest of all canids, weighing scarcely 1.5 kg (3.3 lb) (see fig. 4.2C). This xeric-adapted canid inhabits the deserts of northern Africa, Sinai, and Arabia and can maintain water balance for a minimum of 100 days when fed only mice and no drinking water (Noll-Banholzer 1979). It has numerous adaptations for life in xeric environments. The sandy color of its **pelage** assists in camouflaging the fennec in its desert environment, while its thick, lush coat provides **thermoregulatory** advantages during the cold nights in the deserts. The soles of the feet are covered by long, soft hairs completely concealing the pads—an adaptation for walking on hot sandy substrates, as well as aiding locomotion in loose sand (Larivière 2002).

Evaporative Cooling

Metabolic processes, including the kidney functions mentioned above, require energy. Changes in metabolic processes produce heat, and in desert environments, internal heat must be lost or an individual overheats and dies. Small desert mammals cannot remain exposed to summer conditions for lengthy periods without resorting to **evaporative cooling** for heat dissipation. If they used only the same heat-loss mechanisms as larger mammals, their larger ratio of surface area to body mass would force them to lose large volumes of water, which would quickly cause dehydration. Small mammals do not possess sweat glands, and cutaneous water loss occurs by diffusion through tissues. Evaporative cooling is relatively simple. When mammals cool by evaporation, they take advantage of a physical property of water: its ability to absorb a great deal of heat as it changes state from liquid to vapor. Compared with mesic-adapted species, small desert mammals are characterized by low evaporative water loss. This is the result of a reduction in both cutaneous and respiratory water losses. In desert ecosystems, however, heat is intense and water scarce, so evaporative cooling is of limited utility except as a short-term response to a temperature crisis. In terms of thermal stress, it is clearly maladaptive for a kangaroo rat to venture into the desert sun. For such a small mammal to maintain normal body temperature under such circumstances, it would have to evaporate 13% of its body water per hour. This would be highly taxing, as most species die when they lose 10% to 20% of their body water. As we know from studies of how mammals cope with cold (see Chapter 8), many factors can modify the direct influence of environmental stressors. Although evaporative cooling requires some trade-offs, it represents a major line of defense for mammals in combating heat.

I focus here on four major mechanisms of water loss known as **insensible water loss**, or **transpirational water loss**, which occurs by diffusion through the skin and from the surfaces of the respiratory tract. These mechanisms are sweating, panting, saliva spreading, and respiratory heat exchange (R. W. Hill and Wyse 1989). Keep in mind that small mammals are limited by body size in the extent to which they can store and lose heat. Selection of cool and saturated microclimates is a crucial strategy for conserving water in the small mammals of xeric ecosystems.

Sweating, Panting, and Saliva Spreading

For many mammals, water loss occurs through the skin by way of sweat glands. There are two types of sweat glands: **apocrine** (on the palms of the hands and soles of the feet), which do not secrete for thermoregulation, and **eccrine** (distributed throughout the body), which secrete for evaporative heat loss. Water released from eccrine sweat glands evaporates from the surface of the skin, cooling it and the underlying blood. Sweating in response to overheating occurs only in primates and several species of ungulates; it does not occur in rodents and lagomorphs. For humans working in a hot, dry environment, as much as 2,000 mL/hour of water may be produced by eccrine sweat glands and lost by evaporation. Sweating appears to have evolved in mammals whose fur does not present an appreciable barrier to surface evaporation, but the actual mechanism responsible for evaporative heat loss is not fully known.

Humans sweat to increase cooling by evaporation. In contrast, canids have very few sweat glands and cool primarily by **panting**—a rapid, shallow breathing that increases evaporation of water from the upper respiratory tract. Panting is a common method of evaporative cooling for many carnivores and smaller ungulates, such as sheep, goats, and many small gazelles. Except for some of the small carnivores, of course, the latter species are certainly not small mammals; however, the concepts are crucial to our discussion of regulation in hot environments. All mammals lose some heat as a result of evaporation of water from their respiratory passages. Inspired air is cooler and less humid than expired air, and thus heat is released from the evaporatory surface in both warming and humidifying the air. Water (and heat) is conserved during expiration when the warmed, moist expired air meets the cooler respiratory surfaces.

A major difference between sweating and panting is that the panting animal provides its own air flow over the moist surfaces, thus controlling the degree of evaporative cooling. A sweating animal has minimal control over the degree of evaporation. Another shortcoming of sweating is that sweat contains large amounts of salt. A profusely sweating human may lose enough salt in the sweat to become salt-deficient. This is why we are reminded to drink lots of liquid and limit strenuous exercise outside on very warm days. In contrast, panting animals do not lose any electrolytes and do not become sodium-stressed. Panting does have some drawbacks, however. First, the muscular energy associated with panting generates more heat than does sweating, thus adding to the **heat load**. Second, the increased ventilation generated by panting can result in severe respiratory alkalosis—an elevation of serum pH attributable to excess removal of carbon dioxide.

Panting has the major advantage of allowing an animal under sudden heat stress to maintain a high body temperature and yet keep its brain at a lower temperature. C. R. Taylor (1972) and Mitchell et al. (1997) described this fascinating adaptation in artiodactyls. This concept is best illustrated in large mammals. The brain is kept cooler than the body by the mechanism of **countercurrent heat exchange**. Arteries carrying warm blood from the heart toward the brain come into intimate contact with venous blood cooled by evaporation of water from the walls of the nasal passages in the **cavernous sinus**—a network of small vessels immersed in cool venous blood located in the floor of the cranial cavity where heat exchange occurs. The venous blood from the nasal passages cools the warmer arterial blood heading toward the brain. As a result, the temperature of the brain may be 2°C or 3°C lower than that of the blood in the core of the body. C. R. Taylor and Lyman (1972) found that the small Thomson's gazelle of East Africa, running for five minutes at a speed of 40 km/hour (almost 25 miles/hour), exhibited a core body temperature of 44°C (111.2°F), but its brain was maintained at the cooler level of 41°C (105.8°F). Other physiological devices may augment the cooling of the brain achieved by panting.

As mentioned above, small mammals do not have sweat glands, and cutaneous water loss occurs by diffusion through tissues. Emergency thermoregulation is achieved by salivating on the neck and chest region. When faced with heat stress, many rodents and marsupials spread saliva on their limbs, tail, chest, or other body parts. Grooming saliva assists in evaporative heat loss. This technique is less effective than sweating for evaporative cooling, because the fur must be soaked with saliva before heat can be lost from the underlying skin. Furthermore, this technique is only effective for a short time. Because of their small size, most rodents have limited supplies of internal water to replenish the high rates of loss. Nonetheless, many rodents rely solely on saliva spreading for evaporative cooling. This mechanism is especially useful when heat stress is of relatively short duration—for example, to prevent excessive hyperthermia while searching for cool underground refuge sites.

Secretion of saliva, like sweating and panting, is controlled by the hypothalamus. An increase in body temperature from a **set point** between 37°C and 38.5°C (98.6°F and 101.3°F) is detected in preoptic tissues of the anterior hypothalamus and results in activation of a salivary control center in the brainstem. Saliva then flows from the **submaxillary** and parotid glands (the salivary glands); saliva spreading thereby decreases or stabilizes the rising body temperature, which, in turn, signals the hypothalamus by a negative feedback mechanism. Certain nondesert rodents have been shown to increase evaporative heat loss to more than 100% of heat production, and at least half of this evaporative cooling is due to saliva spreading.

Respiratory Heat Exchange

Heat recovery (heat exchange) in the respiratory system is slightly different from the countercurrent heat exchange process for the vascular system. For example, during inhalation, heat and water are added to air before it reaches the lungs. During exhalation, heat is typically lost to the environment through loss of warm air and evaporation of water. In the vascular system, heat exchange occurs at the same time between two channels, but in the respiratory system, the exchange occurs at different times in the same channel (Schmidt-Nielsen 1997). Heat exchange in caribou demonstrates the principle. When active, caribou produce a considerable amount of heat—which, due to their excellent insulation, may exceed their thermoregulatory requirements. Some heat must therefore be lost, otherwise their body temperature would increase. When resting, caribou do not generate as much heat and must minimize heat loss to the environment. Caribou and some other large mammals have a flap of skin at the opening to each nasal passage. This flap reduces the opening to a small slit when the animal is inactive, thus minimizing heat loss from the lungs. When resting, the temperature of expired air is lower than body temperature. When exercising, the nostrils flare open, increasing the surface area of the mucous lining of the nose for heat exchange. Due to the increased rate of respiration during activity, the volume of air traveling through the nasal passages has less time for heat exchange. The temperature of expired air during exercise is therefore higher when the animal's heat production exceeds its requirements for temperature regulation. Thus, respiratory heat exchange for large arctic species serves a dual function: minimizing heat loss while at rest and increasing heat loss during periods of activity (Scholander 1957; Hainsworth 1981: 223).

Many desert mammals, such as kangaroo rats (*Dipodomys*) (fig. 9.1) are able to cool expired air and thus reduce the amount of water lost to the environment. Actually, much of the exhaled water is recovered by countercurrent heat exchange in the nasal passages. The nasal cavity has a series of thin, convoluted sheets of bone, called turbinals. The turbinals, covered by olfactory epithelium, greatly increase the tissue surface area. As air is inhaled, it passes over these moist tissues in the nasal passages and is warmed and humidified. As the relatively dry air passes over the moist tissues of the nasal passages, these tissues are cooled by evaporation, and heat is transferred from the tissues to warm the inhaled air. During exhalation, the returning warm, saturated air from the lungs condenses on the cool walls of the nasal passages, thus conserving water. This countercurrent exchange—evaporation on inhalation and condensation on exhalation—conserves both water and energy. Kangaroo rats are extremely efficient at cooling expired air, because of the unique morphology of their nasal passages. Compared with another desert inhabitant, the cactus wren (*Campylorhynchus brunneicapillus*), the nasal passages of kangaroo rats are very narrow with a large wall surface, which enhances heat exchange between the air and the

FIGURE 9.1. Merriam's kangaroo rat (*Dipodomys merriami*). Desert-adapted species of rodents such as Merriam's kangaroo rat are able to reduce evaporative water losses by nasal heat exchange and/or by living in humid microenvironments—namely, subterranean burrows. Living in such saturated microhabitats helps to reduce the vapor pressure deficit between the evaporating surface and the surrounding air. See Feldhamer et al. (2007: fig. 9.22) for a depiction of the comparative anatomy of the nasal passages of cactus wrens and kangaroo rats. *Source:* T. L. Best.

nasal tissues. In the cactus wren, the passageway for air flow is wider and shorter and thus the surface area for contact is smaller. As a result, the nasal passages are less efficient at heat exchange than those of the kangaroo rat.

Pulmocutaneous evaporation—loss of water from the respiratory passages and through the skin—is a major avenue of water loss for small mammals in dry habitats (French 1993). For example, of the total water lost by *D. merriami,* 84% is lost by evaporation in the respiratory tract and 16% is lost through the skin. The orange leaf-nosed bat (*Rhinonycteris aurantius*) is a tropical cave-dwelling bat of northern Australia. The species has a strong preference for hot and humid roosts (28°C–32°C and 85%–100% relative humidity). Its rate of pulmocutaneous evaporation is more than double that of other bats and about seven times that in rodents of similar body mass (Baudinette et al. 2000; Tracy and Walsberg 2001). Desert-adapted species of rodents and many species of bats are able to reduce evaporative losses either through nasal heat exchange (in rodents) or by inhabiting humid microenvironments such as **subterranean** burrows and caves (in bats). Living in water-vapor-saturated microhabitats helps reduce the vapor pressure deficit between the evaporating surface and the surrounding air.

We commonly think of pelage insulation as an important mechanism used by cold-adapted mammals to minimize the flow of heat from the core of the body to the environment. However, pelage is also important as an adaptation for xeric-adapted species of small mammals. Studies of the insulation of small desert rodents indicate a trend toward increased insulation. McNab and Morrison (1963) examined *Peromyscus* from arid and mesic environments and concluded that desert subspecies showed a marked increase in insulation compared with their nondesert relatives. When the body temperature is less than the ambient temperature (e.g., during summer in desert ecosystems), it is advantageous to have fur of low conductance so as to slow the inward conduction of heat. The body temperature of desert rodents decreases at night; during the day, heat is gradually stored in the fur, with a concomitant rise in body temperature.

Additionally, the pelage of many mammals adapted to hot environments is not uniformly distributed over the body. Gradation in pelage is best illustrated in large mammals. For example, guanacos (*Lama guanacoe*), living in regions of South America characterized by intense solar radiation and high air temperatures, have gradations in pelage ranging from bare skin at the axilla, groin, scrotum, and **mammae** (mammary glands) to thick pelage on the dorsum. These bare and sparsely furred areas, seen also in many desert antelopes, serve as **thermal windows** through which some of the heat gained from solar radiation can be lost by convection and conduction (Morrison 1966). For amphibious pinnipeds, dissipation of heat is crucial when exposed to external thermal stress on land. These mammals have thermal windows distributed throughout their body surface (Øritsland, Lentfer, and Ronald 1974; Mauck et al. 2003; Willis et al. 2005), which act as efficient dissipaters of heat during thermal stress. In **terrestrial** mammals such as guanacos, African elephants (*Loxodonta africana*), and woodchucks (*Marmota monax*) (Phillips and Heath 1992, 2001), thermal windows typically are bare or sparsely haired thermoregulatory surfaces mainly located in the appendages, which permit heat loss by increased peripheral blood flow (T. M. Williams 1990; Klir and Heath 1992).

Woodchucks (*M. monax*) are well adapted to cope with increased ambient temperatures encountered during their diurnal foraging bouts (fig. 9.2A). Woodchucks use vasodilation of their feet, ears (**pinnae**), and nose to

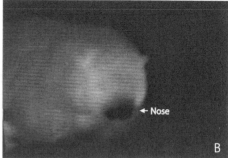

FIGURE 9.2. (A) The woodchuck (*Marmota monax*). (B) Thermogram of a woodchuck. At an ambient temperature of 30°C (86°F), the temperature of the bare nose surface is 27.5°C (81.5°F), or 2.5°C lower than the ambient temperature. The mean temperature for the remainder of the head is 32°C (89.6°F). *Sources:* (A) Hal S. Korber; (B) adapted from Polly K. Phillips, in P. K. Phillips and J. E. Heath (*Comparative Biochemistry and Physiology.* 129A:557–562, 2001).

facilitate heat loss at high temperatures. Using thermographic techniques, Phillips and Heath (2001) demonstrated that at an ambient temperature of 30°C (86°F), the mean temperature of the bare nose surface of *M. monax* was 2.5°C lower than ambient (fig. 9.2B). The bare nose surface combined with long, narrow air passages may offer a type of countercurrent cooling system (see Chapter 9) that prevents increased brain temperatures during foraging in hot environments.

Behavioral Avoidance of Heat

Small mammals avoid extreme temperatures in desert ecosystems by adhering to fairly definite periods of activity. With the exception of ground squirrels and chipmunks, all desert rodents of North America are nocturnal and **fossorial**. Small mammals optimize survival by avoiding extremes in temperature, which is achieved by staying in burrows or crevices belowground during the heat of the day. For a typical desert ecosystem, the temperature of the air in the burrow is mild compared with the extremes on the surface. Remaining in a burrow system reduces evaporative water loss immediately. By employing **estivation** to decrease body temperature, small mammals can accrue additional energy savings. Together, these behavioral and physiological tactics can reduce water loss to less than 5% of that incurred at normal body temperature aboveground. **Granivorous** desert rodents search for seeds at night when it is cool and spend the day in relatively cool, humid burrows. They are seldom exposed to air with high temperatures. Within burrow environments, the air is saturated with water vapor. This is essential for granivorous rodents, for two reasons: first, the seeds gathered on the dry surface absorb water when stored in the burrow, and second, evaporative water loss is reduced considerably while the rodent is in its burrow because there is no saturation deficit.

Unlike nocturnal heteromyids, small dasyurids and diurnal desert rodents must devise other ways to cope with heat stress. Mark Chappell and George Bartholomew (1981) studied antelope ground squirrels (*Ammospermophilus leucurus*) in the deserts of southern California and elucidated some unique tactics used by these squirrels for surviving heat stress. They found that ground squirrels foraging on the surface at midday in summer were faced with temperatures approaching 75°C (167°F). Daily foraging therefore assumed a bimodal activity pattern, with most activity taking place during midmorning and late afternoon. During the day, body temperature was quite labile, varying from 36.1°C to 43.6°C (97°F–110.5°F). Squirrels used fluctuating body temperature to store heat during their periods of activity. High temperatures limit the time that squirrels can be active in the open to no more than 9 to 13 minutes. During this time, squirrels moved rapidly from one patch of shade to the next, pausing only to seize food or monitor predators. Exposure to midday temperatures was minimized by running or "shuttling" between foraging and cooling sites. The body temperature showed a pattern of rapid oscillations, rising while the squirrel was in the sun and dropping when it retreated to its burrow.

The Cape ground squirrel (*Xerus inauris*) lives in the hot, arid regions of southern Africa. These ground squirrels, about the size of the North American gray squirrel (*Sciurus carolinensis*), are found in open woodlands, grasslands, and rocky country where temperatures may reach up to 43°C (109°F). One of the most conspicuous anatomical features of these ground squirrels is a large, bushy tail that is dorsoventrally flattened and nearly as long as the body. Bennett and coworkers (1984) studied the role of the tail as a parasol, or heat shield, in the Cape ground squirrel. The squirrel holds its wide, flat tail tightly over its back, with the white ventral surface upward (fig. 9.3). In this position, the tail shades a large portion of the animal's back, thus lowering the effects of solar radiation on body temperature. On a hot day, to maintain their body temperature below 43°C,

FIGURE 9.3. The Cape ground squirrel (*Xerus inauris*) of southern Africa endures midday temperatures exceeding 37°C (98.6°F). One strategy used by this squirrel to deal with heat stress is to turn its back toward the sun and use its tail for shade. This "parasol tail" can reduce the temperature of the squirrel by up to 5°C. *Source:* Jane M. Waterman.

squirrels must retreat to burrows every few minutes. Burrows deeper than 60 cm (23 inches) usually have temperatures between 30°C and 32°C (86°F–89.6°F).

Employing one's tail as a heat shield is a fascinating adaptation to cope with the intense solar radiation of desert environments; however, Cape ground squirrels also show other unusual life history traits. As with North American prairie dogs (*Cynomys*), Cape ground squirrels are highly social and reside in colonies. Such colonies typically number 6 to 10 individuals, but some may serve as home to 30 individuals. Ground squirrels are reported to share burrows with two species of mongoose, the suricate (*Suricata suricatta*) and the yellow mongoose (*Cynictis pencillatus*). It is possible that suricates and ground squirrels each benefit from the burrowing activities of the other, and ground squirrels may also benefit from the antipredator behaviors of the suricates and yellow mongooses.

Cape ground squirrels have a social system unlike that of any other mammal—two independent social systems, male and female, coexist (Waterman 1995). Females live in groups of kin (related females and their young) and are cooperative breeders (they help raise each other's young). Males live in amicable all-male social groups that persist throughout the year. Such permanent all-male groups are rare in mammals. What is the adaptive significance of this unique social system? Jane Waterman and coworkers have found that the greatest benefit these animals derive from living in such groups is an enhanced ability to detect (through vigilance) and deter (through mobbing or attacking) predators (Waterman and Roth 2007). Improved survival through group living may offset the difficulties of reproducing in the harsh desert of South Africa. Cape ground squirrels breed throughout the year, even in the winter when there is no precipitation and foraging can mean moving far from their burrows.

Waterman and colleagues also found that up to 70% of female Cape ground squirrels that attempt to breed are unsuccessful. This failure to produce offspring seems to be influenced by experience (females breeding for the first time are less successful than females who have already bred successfully) and by parasite loads; however, reproductive success is not limited by food resources. This latter finding was unexpected because availability of resources supposedly has the greatest impact on females, especially during **lactation**; however, Waterman and colleagues did not find any influence on breeding success, offspring size, or litter size when they compared two sites with very different rainfall and food supply. The researchers did find that parasites had a huge impact on reproduction. When parasites were removed from some female groups and not others (the controls), a fourfold increase in reproductive success was found in the groups without parasites compared with control groups. Contrary to what was expected, removal of parasites led to an increase in energy needs (as determined by resting metabolic rate).

CHAPTER TEN

Ecogeographic Rules

How do variations in **metabolism,** body dimensions, and coloration of mammals relate to their geographic distribution—or do they? Several ecogeographic rules have been proposed to explain the morphological variation of animals on a geographic scale.

Modified Size of Appendages (Allen's Rule)

Besides the mechanisms discussed in Chapter 8, conservation of heat in species occupying cold regions can also be enhanced by reducing the length of exposed extremities. According to **Allen's rule,** the appendages of an **endothermic** species living in colder climates are shorter than those of the same species found in warmer climates (J. A. Allen 1877). By reducing unnecessary body surface area, animals can reduce heat loss. It seems logical from the standpoint of thermal physiology that a long nose and tail, as well as long ears and legs, tend to add unnecessarily to the body surface and would be great heat wasters for cold-adapted species (Scholander 1955). For instance, arctic foxes (*Vulpes lagopus*) have small, rounded, densely furred ears, a reduced muzzle, and short, stubby legs. Foxes living in deserts, such as kit foxes (*V. macrotis*) of the southwestern United States or the fennec (*V. zerda*) of North Africa, have large, elongated, thinly furred ears and comparatively long legs (fig. 10.1). Species such as the black-tailed jackrabbit (*Lepus californicus*) and the snowshoe hare (*L. americanus*) exhibit a similar relationship between appendage size and environment (see fig. 8.2). For species residing in hot climates, long appendages are a valuable commodity because they aid in the dissipation of heat. In four species of macaques (*Macaca*, family Cercopithecidae) of Asia, relative tail length

FIGURE 10.1. The ears of foxes help in regulating body heat. (A) The kit fox (*Vulpes macrotis*) of the American southwest has relatively long ears that help in dissipating heat. (B) The red fox (*V. vulpes*) of temperate America has intermediate-sized ears. (C) In contrast, the relatively short ears of the arctic fox (*V. lagopus*) help to conserve heat in the cold. *Sources:* (A) D. G. Huckaby, ASM image library; (B) Raimund Linke/Photoshot; (C) R. Riewe, ASM image library.

generally decreases with increasing latitude, in accord with Allen's rule (Fooden 1997; Fooden and Albrecht 1999). However, as with Bergmann's rule (discussed later in the chapter), the validity of this generalization is debatable and there are exceptions. For example, the body form of bush dogs (*Speothos venaticus*) in the tropics of South America is not unlike that of arctic foxes. Furthermore, the length of ears and tail in hares (*Lepus*) follows Allen's rule, but that for rabbits (*Sylvilagus*) does not (Stevenson 1986).

Seasonal Color Dimorphism (Gloger's Rule)

Although northern mammals exhibit some fascinating anatomical specializations for coping with harsh winter climates, perhaps the most obvious trait defining them is their white **pelage** (fig. 10.2). **Gloger's rule** states that "races in warm and humid areas are more heavily pigmented than those in cool and dry areas" (Gloger 1833; Mayr 1970: 200).

Pigmentation in many mammalian species tends to be paler in habitats closer to the Arctic, with northern races (subspecies) of animals generally being lighter in color than their southern counterparts. Coloration of the pelage of many arctic mammals either remains white year-round or changes to white during winter. Mammals that remain white year-round include the polar bear (actually, its hair is transparent and pigmentless, a property discussed later in the chapter), the arctic hare (*Lepus arcticus*) of the far north, and northerly forms of the caribou and gray wolf. It is noteworthy that arctic hares in the far north, such as those on Ellesmere and Baffin Islands and in Greenland, do not change color seasonally but remain white during the short arctic summer (fig. 10.3). Because hares are so well adapted to winter, the cost in energy of changing to brown during the short summer would exceed the benefits of permanent adaptation to winter. A seasonal color change (dimorphism) is seen in collared lemmings (*Dicrostonyx*), Siberian hamsters (*Phodopus sungorus*), snowshoe hares (*L. americanus*), mountain hares (*Lepus timidus*), ermines and weasels (*Mustela*), and arctic foxes (*V. lagopus*).

Species that undergo seasonal dimorphism accomplish the color changes in different ways. The hairs of ermines and weasels, for example, turn white along their entire length. This complete molt is achieved by each hair being lost, with a new, white (winter) hair replacing the old, brown (summer) hair. The snowshoe hare exhibits a different type of molt. Rather than going through a complete molt pattern, only the tips of the winter hairs of *L. americanus* are white; the bases remain gray. The timing of molt patterns for the Old World mountain hares is most strongly influenced by length of day (photoperiod) but is also correlated with average ambient temperature and duration of snow cover (Angerbjörn and Flux 1995). Furthermore, species of mammals that show seasonal color changes do not change color over

FIGURE 10.2. The short-tailed weasel, or ermine (*Mustela erminea*), undergoes a seasonal color change in northern regions of its geographic range in North America. (A) Summer pelage is brown, and (B) winter pelage is white. The adaptive advantage of seasonal color changes in small mammals is most likely in concealment for both predator and prey. *Sources:* (A) J. A. Wrazen, ASM image library; (B) P. K. Anderson, ASM image librar.

their entire geographic range. For example, the North American long-tailed weasel (*Mustela frenata*) exhibits seasonal color changes north of approximately 40°N latitude. However, south of this zone these weasels remain brown throughout the year. Individuals occupying the zone of overlap between these color groups show gradations in color from white to pied or brown.

The specific cue (the proximate factor) that triggers the onset of molt is decreasing day length. In autumn, the optic nerve receives the stimulus, which is then transmitted to the hypothalamus. The ultimate factor responsible for establishing the duration of the winter coat is probably its thermal advantage, accounting for the close correlation between molting and mean annual temperatures (Flux 1970; Johnson 1984). Temperature and photoperiod, however, are not the sole determinants of white winter coloration. Transplantation experiments (the transfer of animals from one geographic region to another) have demonstrated that heredity as well as temperature is involved in controlling pelage changes in mammals (Kliman and Lynch 1992). The molt cycle is geared to changes in the environment by way of changes in the hormonal system (the endocrine glands: thyroid, pituitary, adrenal cortex, and pineal) as mediated by the brain and the hypothalamus. The seasonal cycles of molt and reproduction are closely related and are coordinated by the neuroendocrine system. In spring, in addition to changes

FIGURE 10.3. Arctic hares (*Lepus arcticus*) live on the arctic tundra—the most northern distribution of all leporids. Most of these hares show seasonal changes in coat color, but arctic hares inhabiting the far north, as shown here, keep their white coats year-round, even though there is no snow cover in midsummer. *Source:* Carnegie Museum of Natural History.

in pelage color, density, and length, the brain signals the pituitary to secrete gonadotropins, which stimulate the gonads to prepare for the approaching breeding season. During spring, the hair follicles and testes of male weasels enlarge simultaneously; the testes begin to manufacture testosterone and sperm, and the hair follicles accumulate melanin and manufacture hair for a complete coat replacement (C. M. King 1989).

Why northern animals turn white in winter is not fully understood. We assume that if the mechanism were not maintained by natural selection (if the mechanism did not confer an **adaptive** advantage more often than a disadvantage), it would disappear. We assume that the color white acts to conceal both predators and prey (Stoner, Bininda-Emonds, and Caro 2003). For example, this adaptation may conceal polar bears from a potential prey animal. While actively searching for prey, however, a weasel is surely easily detected, even with a white pelage. It is possible that the weasel's white color in snowy regions may allow it to blend with its background (cryptic coloration) and thus avoid predation by hawks, owls, and foxes. Many questions concerning the adaptive nature of color changes in northern mammals have yet to be answered (Marchand 1996). For example, what can be the explanation for the blue color phase that commonly occurs in arctic foxes inhabiting the Pribilof Islands and many coastal areas of Alaska and Canada? If color acts to camouflage these animals, why are they slate blue during winter? If this blue phase is not adaptive, how did it evolve and how is it maintained in the population? If cryptic coloration is important to conceal predator from prey, why don't gray foxes (*Urocyon cinereoargenteus*), fishers (*Martes pennanti*), and martens (*Martes*) turn white in snowy regions?

The color white may also convey a thermal advantage. According to the laws of physics, black-colored animals lose heat by radiation faster than white-colored ones. Following Gloger's rule, we would expect to find white animals in cold regions. Some investigators have misinterpreted the pertinent physical laws, however. Radiation of heat from an animal's body is in the form of infrared energy, which is unrelated to visible coloration. All animals are therefore considered to be thermodynamic "blackbodies," meaning that they absorb all incident radiation and reflect none. The color of the fur and underlying skin may be important to the amount of heat absorbed from solar radiation—its peak intensity is in the visible range. When exposed to direct solar radiation, dark-colored skin or fur absorbs more incident energy than light-colored skin or fur. The conservation of this heat energy depends on the length and thickness of the fur, not on its color. As noted above, mammals in northern regions may not only have white fur but also have thicker, denser pelages. Glen Walsberg (1991) examined coat insulation and solar heat gain in three species of subarctic mammals that shift between white winter pelages and darker summer pelages (*L. americanus*, *Mustela erminea*, and *P. sungorus*). He contends that seasonal coat changes in these species serve important roles in cryptic coloration, thermal insulation, and radiative heat gain. A white coat in winter does indeed result in a reduction in solar heat gain; however, this is not necessarily a result of increased heat reflectivity of the pelage or optical properties of the coat in relation to solar radiation; rather, it may be a product of increased coat insulation (Walsberg 1991). Generalizations concerning the adaptive significance of different ecogeographic "rules" or trends must take into account that the survivability of northern mammals may not result solely from forces that maximize heat conservation.

For some animals, the thermal advantage of hair is not necessarily contingent on its color, density, or length but rather on the anatomy of the hair and color of the skin beneath it. For many members of the family Cervidae, insulation depends on the air contained by the highly medullated **guard hairs** that provide insulation (Johnson and Hornby 1980). Research on polar bears indicates that a combination of transparent, pigmentless hair and black skin may enhance their heat-conserving abilities (Grojean, Suusa, and Henry 1980; Walsberg 1983). Pigmentless hair traps and transmits to the skin 90% of sunlight in the invisible ultraviolet portion of the spectrum but only 10% of the visible light. Energy in the form of heat from ultraviolet light is absorbed by the dark skin to help warm the body, while visible light is reflected as white color. The hairs of these large mammals act as optical fibers, with ultraviolet light entering at one end and bouncing along the inside of the hair shaft to reach the dark skin, where it is absorbed.

Body Mass and Latitude (Bergmann's Rule)

How do body mass and metabolism relate to the geographic distribution of mammals? Several ecogeographic rules have been proposed to explain morphological variation on a geographic scale. The best known ecogeographic rule is **Bergmann's rule**. In 1847, Carl Bergmann contended that "on the whole ... larger species live farther north and the smaller ones farther south" (quoted in James 1970). Ernst Mayr (1956: 105) restricted Bergmann's rule to variation within a species, stating that "races of warm-blooded animals from cooler climates tend to be larger than races of the same species from warmer climates." Investigators testing the validity of Bergmann's rule have traditionally compared body size with latitude, using latitude as a proxy for temperature (Ashton, Tracy, and de Queiroz 2000). There is much debate concerning the interpretation and implications of Bergmann's research (Feldhamer et al. 2007).

Bergmann's rule implies that some energetic advantage may be gained through a decreased surface-area-to-volume ratio. We can further generalize that the amount of heat loss depends on both the animal's surface area and the difference in temperature between its body surface and the surroundings. Large mammals, for example, have less heat loss than small mammals because of their larger body mass relative to surface area. To understand this, visualize a mammal as a cube. With linear dimensions of 1 cm on each side, a cube has a surface area of 6 cm^2 and a volume of 1 cm^3; hence, the surface-area-to-volume ratio is 6:1. If we double the linear dimension of the cube, the total surface area increases to 24 cm^2 and the volume to 8 cm^3; now the surface-area-to-volume ratio is 3:1. By doubling the length, heat conservation is enhanced by reducing surface area relative to volume. Granted, mammals are not cubes, but arguments in support of an energetic interpretation of Bergmann's rule follow this logic: a larger mammal has less total surface area per unit volume (and unit mass) and thereby benefits from a reduced rate of cooling. This interpretation sounds convincing, but there are problems. Small mammals know little about per-gram efficiency and are only concerned with total food requirements; a larger mammal clearly requires more food—a clear disadvantage for **herbivores** during winter.

Interpretation of Bergmann's rule can be tricky. If we analyze body size of mammals distributed over a wide latitudinal range, we see mixed results. Rensch (1936) showed that 81% of North American species of mammals and 60% of European species of mammals were indeed larger at higher latitudes. However, McNab was critical of the work of Rensch (1936, 1938) because Rensch's data were derived from a field guide rather than from measurements of individuals from different localities. McNab (1971) found that of 47 North American species examined, only 32% (15 of 47) followed the trend predicted by Bergmann's rule, which was thus invalid. Geist (1987) concurred, indicating that Bergmann's rule was invalid: although body size of large mammals initially increased with latitude, it reversed between 53°N and 63°N; small body size occurs at the lowest and highest latitudes. However, Ashton and colleagues isolated problems with McNab's analysis and found broad support for Bergmann's rule for all orders and most families of mammals examined (78 of 110 species examined) (Ashton, Tracy, and Queiroz 2000: tables 1–3). Temperature and latitude were strong predictors of body size variation (Ashton, Tracy, and Queiroz 2000; Freckleton, Harvey, and Pagel 2003). In sum, an analysis of the validity of Bergmann's rule must integrate many factors associated with the biology of a species, such as pelage, behavior, temperature- and water-related factors, size and type of food, primary plant production, and morphology, to mention just a few (Rosenzweig 1968; McNab 1971; Burnett 1983; Steudel, Porter, and Sher 1994).

Some species of small mammals conform to Bergmann's rule while others do not. For example, external and cranial dimensions of California mice (*P. californicus*) increase in a stepped cline (gradient) from 30° 29′N to 38°N latitude (Grinnell and Orr 1934; Merritt 1978). Gray squirrels (*Sciurus carolinensis*) exhibit their greatest body mass in the north and smallest in the south (Havera and Nixon 1978). However, fox squirrels (*S. niger*) show a clinal variation in body size in reverse of Bergmann's rule east of the Appalachian Mountains—body size is smallest in the north; fox squirrels west of the Appalachians, however, do follow a pattern predicted by Bergmann's rule (Weigl et al. 1989).

Carnivorous small mammals pose an interesting case when considering energy requirements. The long, thin shape of weasels (*Mustela*) has distinct disadvantages in cold climates. This shape exposes a large surface area to the cold air, and weasels' unique feeding requirements dictate that they pursue prey through small crevices and a labyrinth of **subterranean** runways, which is energetically very expensive. Their mobility would be greatly compromised by a dense, heavy pelage; thus, a short fur is essential for optimal agility. Compared with woodrats (*Neotoma*), weasels have about a 15% higher surface-area-to-mass ratio—they cannot assume a spherical form by curling as woodrats can. As a result, cold-stressed weasels have **metabolic rates** 50% to 100% greater than those of less slender mammals of comparable size. So, being long and skinny may facilitate capturing prey, but weasels pay a high energy cost in the form of a rapid rate of heat loss in cold conditions. As a result, weasels must consume more food to maintain energy efficiency (J. H. Brown and Lasiewski 1972; C. M. King 1989, 1990; McNab 1989).

CASE STUDIES
....

Communal Nesting

Southern flying squirrels (*Glaucomys volans*) are the smallest tree squirrels in North America. In northern environments they face severe thermal challenges during **arboreal** foraging bouts. Because of their small mass, large surface-to-volume ratio, and limited capacity for increasing pelage insulation, group nesting is highly advantageous for energy conservation during the long, cold winters.

Southern flying squirrels, also known as fairy diddles, are abundant in the mixed deciduous forests of the Appalachian Mountains. They were a common inhabitant of my study site in a beech-maple-poplar community at Powdermill Biological Station in western Pennsylvania (fig. 10.4A). *G. volans* showed high survival during winter—it does not **hibernate** or even drop body temperature during winter months. Flying squirrels establish globular nests of shredded bark, leaves, and grasses in tree cavities such as abandoned woodpecker holes. These nests are occupied by groups of squirrels known to form "huddles" composed of up to 20 individuals; however, 10 to 14 was the usual number on my study site (fig. 10.4B).

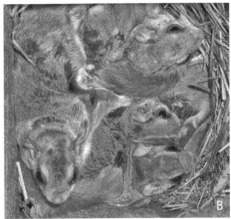

FIGURE 10.4. (A) The southern flying squirrel (*Glaucomys volans*). (B) These flying squirrels commonly establish communal nests of 10 to 14 "flyers" per nest in cavities of trees as a way to cope with winter cold. Southern flying squirrels were studied in an outdoor enclosure (C) and an outdoor laboratory to determine the role of communal nesting in enhancing overwinter survival. *Sources:* (A) Hal S. Korber; (B, C) J. F. Merritt.

The objective of our study was to evaluate two mechanisms adaptive in enhancing the overwinter survival of *G. volans*. This was done, first, by evaluating monthly changes in energy requirements (using oxygen analysis) for squirrels residing singly in nests compared with those nesting in aggregations. We predicted (logically) that squirrels nesting in a group would conserve energy by exhibiting a reduced metabolism compared with squirrels nesting singly. Second, we analyzed monthly changes in the heat-producing capacity of **brown adipose tissue** in the seasonal heat production of squirrels nesting singly compared with those nesting in groups. We hypothesized that flying squirrels maintain the high body temperature necessary for **nocturnal** foraging during periods of cold by using huddling thermogenesis augmented by increasing capacity of nonshivering heat production (**nonshivering thermogenesis**) by brown adipose tissue.

The experimental design was straightforward. We used an outdoor enclosure (fig. 10.4C) and outdoor laboratory to study seasonal thermogenesis of squirrels nesting singly. We placed flying squirrels singly in a glass terrarium supplied with a wooden nest box, shavings, and hay for bedding. Food (sunflower seeds, acorns, and mushrooms) and water were supplied *ad libitum*. To study the role of group nesting in seasonal energy expenditure, we confined squirrels to an outdoor enclosure that simulated the natural environment. A large section of a hollow sycamore tree was positioned in the middle of the enclosure; this had a central cavity for nesting and **caching**. We provided hay, leaves, and grape vines for nesting material; food in the form of sunflower seeds, acorns, and mushrooms; and water *ad libitum*. Apples and oranges were provided to both sites on holidays! Animals were monitored once a day for the year-round study in both the outdoor laboratory and outdoor enclosure. Metabolic tests were conducted in the nearby laboratory for five-day periods once a month. The squirrels were returned to the nesting sites immediately following the metabolic runs.

The results of the study were fairly predictable in the light of the first law of thermodynamics (energy conservation). Flyers in group nests demonstrated a lower metabolic rate during winter than flyers housed singly—those hot bodies in the huddle help to decrease flow of heat from the nest core to the outside environment. The squirrels on the outside of the huddle are not always the losers and will rotate to the center of the huddle periodically. Brown adipose tissue and nonshivering thermogenesis are also important to flyers in winter, as a quick source of heat to combat cold temperatures while foraging on winter nights.

Hedgehogs: Spiny Hibernators

Hedgehogs are members of the family Erinaceidae (fig. 10.5), which includes 10 genera and 24 species. They are characterized by barbless spines on the back and sides. Erinaceids are an Old World family found in Africa,

FIGURE 10.5. (A) Long-eared hedgehog (*Hemiechinus auritus*). Hedgehogs, members of the family Erinaceidae, are well known for the arsenal of barbless spines on their back and sides. Hedgehogs occur in Africa, Europe, and Asia, including Sumatra, Borneo, and the Philippines, where they inhabit many different habitats such as forests, grasslands, fields, and farmlands. (B) Southern African hedgehog (*Atelerix frontalis*). Hedgehogs tend to curl into a ball for protection when faced with predators. *Sources:* (A) M. Andera, ASM image library #489; (B) J. Visser, ASM image library.

Europe, and Asia, including Sumatra, Borneo, and the Philippines. They are found in many different habitats, including forests, grasslands, fields, and farmlands. As their name suggests, desert hedgehogs (*Paraechinus*) inhabit arid areas in North Africa, India, and Pakistan. Long-eared desert hedge-hogs (*Hemiechinus*) are found from northern Africa east to the Gobi Desert of Mongolia. Hedgehogs are mainly nocturnal and **terrestrial**, but some are semiarboreal. An adult *Hemiechinus auritus* weighs only 40 to 50 g (1.4–1.7 oz) (fig. 10.5A), whereas the common European hedgehog (*Erinaceus europaeus*) reaches 1,100 g (2.4 lb). All are **omnivorous**, feeding on small vertebrates, eggs, fruits, and carrion in addition to invertebrates.

Although best known for their arsenal of spines, hedgehogs (and some tenrecs) are unique among the "insectivores" in their ability to employ hibernation as a mechanism for energy conservation in response to reduced energy availability and/or cold. During hibernation, body temperature falls from about 35°C (95°F) to between 15°C and 20°C (59°F–68°F); heart rate decreases from about 250 to about 10 beats/minute. Reportedly, the rate of respiration is quite variable, and breathing can entirely cease for up to two hours. Hedgehogs in captivity do not hibernate. Webb and Ellison (1998) describe recent research on the physiology of **thermoregulation** of European hedgehogs (*E. europaeus*)

A European hedgehog has approximately 5,000 spines, each about 2 to

3 cm (1 inch) long with needle-sharp points. Its defensive posture is typical for a mammal with spines or scales. When hedgehogs are threatened, it rolls into a tight ball with the spines directed outward. This structure is aided by paired longitudinal "drawstring" muscles, the panniculus carnosus, on either side of the body. This musculature is more strongly developed on the edges than in the center. When the muscles contract they act like a drawstring around the opening of a bag, forcing the contents deeper into the bag as the string is drawn tighter. When the hedgehog is in the fully curled position, the spines that covered its flanks and top of its head are brought together, resulting in a small hole (the width of one's finger) corresponding to the opening of the "bag." Hedgehogs have an interesting "self-anointing" behavior in which they spread large amounts of foamy saliva on the spines. This may be done as a sexual attractant during the breeding season, to reduce parasites, to clean the spines, or as additional protection from predators (Wroot 1984; Weldon 2004; D'Havé et al. 2005). Several species, including the common European hedgehog, undergo true hibernation throughout the winter—the only "insectivores" that do so. Desert hedgehogs commonly **estivate** (enter a dormant condition in the summer). Litter size is generally between four and six, and two litters a year may be produced. The young are **altricial** (immature). At birth, their short, soft spines have not yet broken through the skin, but the spines quickly grow in length after birth and harden within a few weeks.

PART THREE

Reproduction

Reproduction is the process by which organisms produce new individuals, passing on genetic material and thus maintaining the continuity of the species and of life. All mammals reproduce sexually by way of internal fertilization. Furthermore, all mammals are dioecious: the sexes are separate, and a given individual normally has the sex organs and secondary sexual characteristics of only one sex. Mammals vary greatly in the structure and function of their reproductive systems.

The *monotremes* lay eggs and incubate their young; like most reptiles and birds, monotremes have a cloaca, a common opening for the urinary and reproductive tracts. In addition, as in birds, only the left ovary is functional in monotremes. The monotremes have true mammary glands but lack nipples (Grant 1995; Rismiller and McKelvey 2000). The *marsupials*, or pouched mammals, also have a cloaca. The placenta in these animals is not very efficient, due to the limited contact between the maternal and fetal blood supplies. As a consequence, marsupials undergo a very brief gestation period and a prolonged period of nursing. The young are born in an altricial (poorly developed) state and require a long period of development in the mother's pouch (marsupium) (Padykula and Taylor 1982; Dawson 1995; Tyndale-Biscoe 2005).

The eutherian mammals, commonly called *placental* mammals, have made a major advance in the evolution of their reproductive patterns. The key innovation is a highly sophisticated placenta that facilitates efficient respiratory and excretory exchange between the maternal and fetal circulations. Compared with the marsupials, the eutherian mammals have a longer gestation period and a shorter period of nursing or lactation. Furthermore, the young are more highly developed (precocial) at birth than are young marsupials. Mammals vary widely in the complexity of their reproductive patterns.

Despite the differences in reproductive complexity, it would be incorrect to interpret the seemingly more primitive reproductive patterns of the monotremes and marsupials as being less successful than those of eutherians. The success of a species can be evaluated only case by case and in the context of its specific ecological relationships. There are many examples of successful groups of non-eutherian mammals. For example, the success of the North American opossum is demonstrated by its broad geographic distribution and high reproductive potential. This marsupial originated in South America and has colonized environments as far north as Canada within only a short time. Indeed, North American opossums reach as far as southern Ontario; however, their tails and ears are a bit ragged—a product of frostbite.

The variations in reproductive patterns of mammals seem endless. For example, gestation ranges from 10 to 14 days in some dasyurid marsupials to more than 650 days in elephants. Lactation varies from only 4 days in pinnipeds such as the hooded seal (*Cystophora cristata*) to more than 900 days in chimpanzees and orangutans (*Pan troglodytes* and *Pongo pygmaeus*). Most mammals give birth to between 1 and 15 young per litter, and intervals between births range from three to four weeks in rodents to as long as three to four years or more in sirenians, elephants, and rhinoceroses (Hayssen, van Tienhoven, and van Tienhoven 1993). Voles (order Rodentia, family Cricetidae) may hold the record for biotic potential: the meadow vole (*Microtus pennsylvanicus*) of the North American grasslands may produce up to eight or nine litters of five to eight young in a single year. As a result, a single female may potentially produce up to 72 offspring during one breeding season! Bailey (1924) reported that a captive female meadow vole produced 17 families in a single year. Naked mole-rats (*Heterocephalus glaber*, family Bathyergidae) of Africa have the largest mean litter size of all wild mammals: up to 28 young (Sherman, Braude, and Jarvis 1999). Variation in reproductive patterns may be explained by assessing environmental factors and the risk of raising young in highly seasonal environments. Hayssen and coauthors present a thorough review of the seemingly endless reproductive strategies employed by mammals (Hayssen, van Tienhoven, and van Tienhoven 1993).

CHAPTER ELEVEN

Reproductive Variations

For most mammals, fertilization occurs within several hours of insemination. The fertilized egg becomes implanted in the uterus, and development continues until birth. Thus, we define the **gestation period** as the interval between fertilization and parturition. Mammals may exhibit several modifications that lengthen the fertilization or gestation period, including delayed fertilization, delayed development, delayed implantation, and embryonic diapause (Gaisler 1979; Nowak 1999).

Delayed Fertilization

Many mammals inhabiting seasonal environments have special mechanisms to optimize survival. Bats of the families Vespertilionidae and Rhinolophidae that occupy temperate zones of the New and Old Worlds exhibit a reproductive tactic called delayed fertilization, or delayed ovulation (see Feldhamer et al. 2007: fig. 10.7). Copulation takes place in September or October, before **hibernation** commences. Follicular growth has occurred in the ovary, but ovulation does not happen at this time. The sperm are immotile and stored in either the uterus or the upper vagina, and then both sexes enter hibernation. When the female emerges from hibernation in the spring, the eggs are ovulated, spermatozoa become motile, and fertilization takes place. (It is noteworthy that some bats copulate during hibernation, especially species of *Eptesicus, Myotis,* and *Plecotus.*) **Implantation** of the blastocyst (embryo at the 70- to 100-cell stage) occurs about a week after fertilization. The gestation period may last 50 to 60 days, as in silver-haired bats (*Lasionycteris noctivagans*) and little brown bats (*Myotis lucifugus*). Young are born in early summer, when insects are abundant. Delayed fertilization is advantageous for northern species of bats because (1) it provides young with more time to build critical body mass to sustain their long period of hibernation, and (2) it gives adult females adequate time between autumn and spring for all individuals to mate. The latter is particularly important in bats, given their low reproductive rate (one young per year for most species).

Delayed fertilization is most common in temperate species of bats that undergo hibernation; however, mammalogists have found sperm storage in bats living in tropical regions that apparently do not undergo true hibernation. Tropical species may store sperm as an **adaptive** strategy synchronized with the availability of food (Racey 1982).

Little brown bats are among the species that mate in early autumn, prior to hibernation, and the sperm is stored in the uterus during winter. Mating in these bats is also known to occur in winter during the hibernation period. Fertilization occurs in spring when the female bat emerges from hibernation. At this time, females disperse to warmer quarters such as buildings with hot attics, where they establish maternity colonies. Such sites may be occupied by several thousand females. Following the period of hibernation, males venture out and lead bachelor lives throughout summer. It is not known where males spend the summer, but they probably live a solitary life in various types of roosts.

After a gestation period of 50 to 60 days, a single, black, naked, blind offspring is born, usually in late May or early June. During parturition, the female hangs by her thumbs (which is upside down for a bat). She forms a "basket" with her upturned **uropatagium**. This "basket" is used to catch the offspring following the half-hour labor. At birth, a young bat weighs about 1.5 g (0.05 oz). Newborn bats weigh about 20% to 25% of the mother's mass after birth—this is roughly equivalent to a 100 lb human female giving birth to a 25 lb child! Normally, the offspring is left to cling to the roost when the mother forages. If surprised in the roost, however, the mother takes the young bat into flight. She does this by carrying it in crosswise fashion with the infant attached to one nipple, with its hind legs tucked under the mother's opposite armpit. The newborn grows rapidly and begins to fly in the roost at about 3 weeks of age. It begins to forage for insects at about 4 weeks of age, when the colony is disbanded. At about 8 months of age, little brown bats reach sexual maturity. Their average lifespan is about 12 to 20 years; maximum life span is 40 years.

Delayed implantation is a reproductive variation in which ovulation, copulation, fertilization, and early cleavage of the zygote up to the blastocyst stage occur normally. However, development of the blastocyst is arrested, and each blastocyst floats freely in suspended animation in the reproductive tract until environmental conditions become favorable for implantation. Unlike delayed development (discussed below), the blastocyst does not implant in the uterine wall but floats in the reproductive tract. During this free-floating stage, the blastocyst is encased in a protective coat (the zona pellucida) until the optimal time for its development. Eventually, implantation occurs and development proceeds normally. Delayed implantation is either obligate, as in armadillos, in which the delay occurs as a normal, consistent part of the reproductive cycle, or facultative, as when rodents or "insectivores" are nursing a large litter or are faced with extreme environmental conditions. Daniel (1970) and Mead (1989) provide detailed discussions of delayed implantation. Although delayed fertilization and delayed development occur only in bats, delayed implantation is seen in many diverse mammalian groups, including "insectivores," rodents, bears, mustelids (weasels and allies), seals, armadillos, certain bats, and two species of roe deer (*Capreolus*). Delayed implantation is a rule in the pinnipeds (Laws, Baird, and Bryden 2003). Mead (1989) presents an excellent review of delayed implantation in carnivores. Delayed implantation occurs in the Old World vespertilionids *Miniopterus schreibersi* and *M. fraterculus* (Kimura and Uchida 1983). The length of gestation in *M. schreibersi* varies geographically, and differences may in part reflect local environmental conditions (Racey 1982). For many species of bats, timing of implantation may be contingent on endocrine controls for different populations.

Delayed implantation was probably first reported in the mid-1600s, as described in the field notes of William Harvey, who joined King Charles I of England on hunting trips for European roe deer (see Gunderson 1976 for a discussion of the historical account). This reproductive strategy varies widely even within the same genus (Sandell 1984; C. M. King 1990). Siberian and European roe deer (*Capreolus capreolus* and *C. pygargus*) are the only ungulates that exhibit delayed implantation. Following the rut in July and August, implantation is delayed until December or January, and the young are born in the spring (April to June). The gestation period is between 264 and 318 days. In North America, the black bear (*Ursus americanus*) exhibits delayed implantation. In Pennsylvania, for example, black bears mate during the summer, usually between early June and mid-July. After the ova are fertilized, implantation of the blastocyst in the uterine wall is delayed for up to five months. Young are born in winter dens in mid-January, following an "actual" gestation period of about 60 to 70 days. Mustelids also represent excellent examples of delayed implantation (C. King 1983; Mead 1989). Although great variation occurs among different species of the family Mephitidae, the western spotted skunk (*Spilogale gracilis*) seems to represent a fairly typical example of delayed implantation. Female spotted skunks enter **estrus** in September and mate in September or October. The zygotes undergo normal cleavage but stop at the blastocyst stage, at which time they float freely in the uterus for about 6.5 months. After implantation, gestation lasts only about a month, and young are usually born between April and June. The total period of pregnancy is 210 to 260 days. The eastern spotted skunk (*S. putorius*) is also a practitioner of delayed implantation (fig. 11.1). This member of the order Carnivora was aptly described as a "little animated checkerboard" by Ernest Thompson Seton (1929). Eastern spotted skunks, one of the smallest of the New World skunks, are weasel-like in appearance. Their **pelage** is long and soft, jet black, and marked by a triangular white spot on the forehead and a unique pattern of broken white stripes along the back and sides. However, the "checkerboard" coloration and small size are only part of its trademark.

When threatened, the spotted skunk exhibits a unique and highly effective defense behavior—a scare tactic of sorts. Initially, the skunk gives its intruder a warning: its displays an upright tail, erect hairs, and arched back and even stomps its feet. These actions may be accompanied by growls, snarls, or hisses. Next, it employs acrobatics, performing a "handstand" on its forefeet, which is commonly accompanied by foot stomping with its forepaws (Howell 1920). At this point, most predators retreat; those that do not are quickly "educated." *S. putorius* quickly turns its body in a U-shaped position, with head and tail facing the intruder, takes aim, and sprays musk in the victim's face. Discharged from scent glands in the anus, the musk fluid resembles skim milk mixed with small white curds. With the active ingredient of mercaptan, a compound containing sulfur, the strongly acidic spray is quite nauseating and may cause momentary blindness.

Of all mammals, the fisher (*Martes pennanti*) has the most protracted

FIGURE 11.1. The eastern spotted skunk (*Spilogale putorius*) is one of the smallest of the New World skunks. It displays a unique acrobatic feat of performing a "handstand" on its forefeet as a defense behavior. *Source:* Jerry W. Dragoo.

delay in implantation. Fishers breed between March and April. Following a nine-month delay, blastocysts implant in January or February, and most births occur in March and early April (Frost, Krohn, and Wallace 1997). The pregnancy lasts between 327 and 358 days, about the same as that of the blue whale (Hayssen, van Tienhoven, and van Tienhoven 1993).

The adaptive advantage of delayed implantation is poorly understood. Two closely related species of weasels residing in the same habitat may exhibit very different reproductive patterns. For example, in eastern North America, the long-tailed weasel (*Mustela frenata*) and the least weasel (*M. nivalis*) may occupy the same habitat, but the former demonstrates delayed implantation and the latter does not. Many mammalogists believe that by delaying the arrival of the young until spring, when hunting is easiest because small mammals are plentiful, the long-tailed weasel optimizes the likelihood of survival of its young. But why does the smaller least weasel not show such a delay?

The nine-banded armadillo (*Dasypus novemcinctus*) of the New World is another interesting example of variation in reproductive biology. Armadillos in the southern United States breed in July or August. Their copulatory position is rather unusual for a quadruped mammal: the female assumes a mating position on her back. Implantation is delayed until November, followed by a gestation period of about four months. In late February, the litter is born and comprises identical quadruplets, all of one sex—they all come from a single fertilized ovum, a phenomenon called "monozygotic polyembryony." Because only one ovum was fertilized by a single sperm, all the young have exactly the same genetic makeup. Delayed implantation in this case seems to serve to time the birth of the litter for the spring flush of invertebrate food, the staple in the diet of armadillos.

Delayed Development

Delayed development differs from delayed fertilization and implantation in that the blastocyst implants shortly after fertilization, but development is very slow. A fertilized blastocyst implants in the uterus in summer but may have a seven-month period of development, or gestation. Delayed development is reported for both microchiropterans and megachiropterans (Racey 1982). *Artibeus jamaicensis* employs delayed development as a unique mechanism to synchronize the birth of young with the end of the dry season, when the availability of large fruits is at its peak (Fleming 1971; Racey 1982). The Old World **insectivorous** bat *Miniopterus australis* also exhibits delayed development (Richardson 1977). The reproductive delay in this species is reportedly in response to unpredictable availability of insects. Other species of bats reported to undergo delayed development are *Macrotus californicus* (G. V. R. Bradshaw 1962), *Haplonycteris fischeri* (Heideman 1989), *Otopteropus cartilagonodus* (Heideman, Cummings, and Heaney 1993), *Plenochirus jogori* (Heideman and Powell 1998), and *Rhinolophus rouxii, Cynopterus sphinx, Eptesicus furinalis,* and *Myotis albescens* (Racey 1982).

Embryonic Diapause

Many **marsupial** mammals employ a fascinating strategy in reproduction called **embryonic diapause**. This tactic is typified by a blastocyst that enters a state of dormancy, during which time division and growth of the cells may cease or continue at a slow pace until a signal is received from the mother (Tyndale-Biscoe 2005). Embryonic diapause is similar to the process of delayed implantation discussed earlier for **placental** mammals. Among marsupials, embryonic diapause is reported to occur in almost all kangaroos, wallabies, rat kangaroos, pygmy possums, feathertail gliders (*Acrobates pygmaeus*), and honey possums (*Tarsipes rostratus*). Western gray kangaroos (*Macropus fuliginosus*), Lumholtz's tree kangaroos (*Dendrolagus lumholtzi*), and musky rat kangaroos (*Hypsiprymnodon moschatus*) are not known to exhibit embryonic diapause.

An obligatory nectar and pollen consumer, the honey possum (see fig. 3.14) resides in heathlands of the coastal sand plains of southwestern Australia. The reproductive biology of honey possums is fast, furious, and complex. As with the well-known macropods (i.e., most kangaroos, wallabies, and rat kangaroos), honey possums undergo embryonic diapause. Continuous reproduction coupled with embryonic diapause balances the brief, one-year lifespan of these shrewlike mammals. Let's track the role of each parent and the offspring in this fascinating story of reproduction "down under."

The males. Male honey possums are typically smaller than females, and competition for females in estrus is intense. Two attributes are associated with such competition: male honey possums (1) produce the largest sperm of any mammal (0.3 mm, or 0.012 inches, in length)—even longer than that of the blue whale (Renfree, Russell, and Wooller 1984); and (2) have *huge* testes—together with the epididymides, they contribute about 5% of an adult male's body mass. Hugh Tyndale-Biscoe (2005: 191) contends that if human males had testes of this proportion, each would weigh about 2 kg (4.4 lb)!

The females. Female honey possums are **polyestrous** (entering estrus several times a year), and some breed twice per year. Immediately after giving birth, females mate with several males and conceive again. These new embryos develop to the blastocyst stage, but unlike the process in kangaroos, they continue to grow slowly in the uterus for the next three months while the female is **lactating**; they resume full development after the young have left the pouch. The active gestation period is about 21 to 28 days. Honey possums give birth to one to four minute young and each attaches to one of the four teats in the well-developed pouch. Pregnancy is a full-time job with no maternity leave—a female will carry young in her pouch for almost all of her adult life. It is still not clear what controls the period of embryonic diapause during the early stage of lactation. The mechanism is clearly different from that seen in kangaroos, because the pattern is not influenced by experimental removal of pouch young (Tyndale-Biscoe 2005: 191–192). This form of embryonic diapause is very similar to that seen in placental badgers and roe deer.

The young. At birth, young honey possums weigh about 5 mg (0.00017 oz; the smallest known mammalian neonates). They grow and develop in the pouch for about two months, at which time they are furred and weigh about 2.5 g (0.09 oz). Young now leave the pouch and are carried on the mother's back for another month, until they reach 4 to 5 g—almost the size of the mother. Lactation lasts about 10 weeks and is the most demanding period for the mother in terms of providing energy and protein requirements for her rapidly growing young. During this period, a female honey possum may need to visit 2,400 florets in at least eight *Banksia* inflorescences each night to obtain sufficient protein for her young (Tyndale-Biscoe 2005). For the last two weeks of the lactation period, the young are left in a shelter while the mother forages. Young grow rapidly, and in seven to eight months reach sexual maturity. Life is brief for honey possums—they probably live for just one year, and mothers typically produce only about three young during their entire life.

Honey possums are highly dependent on the availability of nectar and pollen produced by eucalypts, *Banksia* and *Callistemon*. Urbanization and habitat modification in southwestern Australia are influencing the honey possums' habitat and limiting the availability of food for this fascinating small mammal.

CHAPTER TWELVE

Mating Systems and Reproductive Strategies

Classification of the mating systems of mammals has traditionally been based on the extent to which males and females associate (bond) during the breeding season. These systems include **monogamy** (a single male and female pair for some period of time and share in the rearing of offspring), **polygamy** (both males and females have several mates), **polygyny** (some males have more than one mate, and females provide most of the care of offspring), **polyandry** (females have more than one mate), and **promiscuity** (both sexes have multiple mates, with no prolonged association between males and females). Small mammals exhibit a wide variety of mating systems and reproductive strategies, many of which I have mentioned in earlier chapters. Many species show fascinating adaptations in reproductive behavior, but some of the most intriguing are the tenrecs, sengis, treeshrews, hammer-headed fruit bats, naked mole-rats, and the brown antechinus. These are the species whose natural histories and reproductive peculiarities I highlight in this chapter.

High Fecundity

When it comes to reproductive competence, tenrecs, inhabitants of Madagascar, are champions in many ways. The tail-less tenrec (*Tenrec ecaudatus*) (fig. 12.1A) may be the most prolific mammal, with litters of up to 32 young, although the average litter is between 15 and 20 (Nicoll and Racey 1985). Tenrecs have 29 teats, which is the maximum known for any mammal. The record for "fastest developer" goes to the lowland streaked tenrec (*Hemicentetes semispinosus*) (fig. 12.1B), whose young are weaned in about

FIGURE 12.1. The tenrecs are a primitive group of insectivorous mammals. Both (A) the tail-less tenrec (*Tenrec ecaudatus*) and (B) the lowland streaked tenrec (*Hemicentetes semispinosus*) are restricted to Madagascar. The common tenrec has 29 teats, the maximum number known for any mammal. The lowland streaked tenrec holds the record for the "fastest developer"—young are weaned in about 5 days and begin to breed as early as 3 to 5 weeks of age. *Sources:* (A) Courtesy of S. C. Bisserot / Bruce Coleman / Photoshot; (B) courtesy of Nigel Dennis / NHPA/Photoshot.

five days and begin breeding as early as 3 to 5 weeks of age (Hayssen, van Tienhoven, and van Tienhoven 1993; Sherman, Braude, and Jarvis 1999). As with many mammalian groups in Madagascar, several species of tenrecs have been reduced in population density and distribution as a result of habitat loss or other factors. The web-footed tenrec (*Limnogale mergulus*) is endangered, as are several other species of tenrecs and otter shrews.

A great diversity of form and function in the order Afrosoricida is reflected in the 30 species of tenrecs and otter shrews, of the family Tenrecidae (Bronner and Jenkins 2005). The 27 tenrecid species are restricted to Madagascar, but the 3 species of otter shrews are found in west-central Africa. Some authorities place otter shrews in a separate family, the Potamogalidae. As suggested by their name, they are semiaquatic and closely resemble river otters (order Carnivora, family Mustelidae). The giant otter shrew (*Potamogale velox*) is the largest living "insectivore," with a total length up to 640 mm (almost 25 inches) and body mass up to 950 g (2 lb) (Kingdon 1997b).

The morphology of tenrecids defies general description, as do their behavior and habitats. Tenrecs are hedgehog-like in appearance, weigh up to about 2.5 kg (5.5 lb), and are relatively common throughout most habitats of Madagascar except for the southwestern arid regions. Their **pelage** consists of hairs and spines, and unlike the greater hedgehog tenrec (*Setifer*), they do not curl into a protective ball. Tenrecs are opportunistic feeders, consuming a variety of invertebrates, some small vertebrates, and plant material. Rice tenrecs (*Oryzorictes*) look like moles, are **fossorial**, and live in marshy areas. Long-tailed tenrecs (*Microgale*) resemble shrews and occupy thick vegetation and ground litter in a variety of habitat types. The web-footed tenrec (*L. mergulus*) has a long, laterally flattened tail and webbed hind feet and looks like a small muskrat. It preys on fish, amphibians, and aquatic invertebrates in rivers, lakes, and marshes. This species is limited to stream habitats in eastern Madagascar and is active only at night (Benstead, Barnes, and Pringle 2001). Earthworms are a staple in the diet of streaked tenrecs (*Hemicentetes*). Several species, such as the streaked tenrec and the long-tailed tenrec, are believed to **echolocate** as part of their foraging activities. The greater hedgehog tenrec (*Setifer setosus*) and small Madagascar "hedgehog" (*Echinops telfairi*), like erinaceids, have sharp, barbed spines on the head, back, and sides. Like hedgehogs, they have well-developed panniculus carnosus muscles and roll into a ball when threatened. Body temperatures of tenrecs are generally low, ranging from 30°C to 35°C (86°F–95°F) while individuals are active. The lowland streaked tenrec (*H. semispinosus*) maintains a body temperature 1°C above ambient temperature while it **hibernates** during much of the winter (P. J. Stephenson and Racey 1994).

As noted above, mating systems can be classified on the basis of the ability of one sex to monopolize or accumulate mates. For instance, individual males in polygynous systems monopolize more than one female, whereas a polyandrous system involves females monopolizing more than one male. "Promiscuity" refers to multiple mating and the absence of any prolonged association by at least one sex. In monogamous systems, neither sex is able to monopolize more than one member of the opposite sex. Generally, monogamy involves a mated pair that remains together for one or more breeding seasons (Kleiman 1977; Reichard and Boesch 2003). This system is rare in the class Mammalia, occurring in less than 5% of all species (J. Clutton-Brock 1989); it occurs among dermopterans, treeshrews, primates, canids, rodents, and some dwarf lemurs and bats. The North American beaver (*Castor canadensis*) has typically been described as a socially monogamous species (Crawford et al. 2008). Other monogamous rodents include California mice (*Peromyscus californicus*), old field mice (*P. polionotus*), prairie voles (*Microtus ochrogaster*), common mole-rats (*Cryptomys hottentotus*), Malagasy giant jumping rats (*Hypogeomys antimena*) (Sommer 1997), and the southern bamboo rat (*Kannabateomys amblyonyx*) (Silva, Vieira, and Izar 2008). In addition, representatives of all four genera of sengis are socially monogamous.

Sengis, or elephant shrews, represent a single family (Macroscelididae), with 4 genera and 17 species, in the order Macroscelidea. Due to their cryptic nature and confinement to Africa, sengis were not described in the scientific literature until the mid-1800s. They are sometimes referred to as "elephant shrews" because of their long legs, long, flexible, highly sensitive snout, and large eyes and ears (fig. 12.2A). To avoid association with true shrews of the family Soricidae, many biologists prefer to use the vernacular name "sengi," which is derived from several Bantu-based languages in central and southern Africa.

Although the sengis' kangaroo-like hind legs give the appearance of a **saltatorial** gait (thus another common name, "jumping shrews"), bipedal locomotion does not occur. Sengis do not jump or hop; rather, the normal method of locomotion is to walk or run on all fours. Size varies greatly— ranging from about 45 g (1.6 oz; the same mass as a meadow vole, *Microtus pennsylvanicus*) in the round-eared sengi (*Macroscelides proboscideus*)

FIGURE 12.2. One of the few socially monogamous mammals is the sengi, or elephant shrew, an inhabitant of deserts, brushlands, and forests of Africa. (A) An adult rufous sengi (*Elephantulus rufescens*). Note the long proboscis, enlarged hind limbs, and kangaroo-like appearance. Evolutionary convergence is evident in the body forms of sengis and several families of rodents (order Rodentia), such as kangaroo rats (family Heteromyidae) and jerboas (Dipodidae). (B) A 1-day-old rufous sengi (*E. rufescens*), illustrating the precocial nature of young sengis. *Source:* Galen Rathbun, California Academy of Sciences.

to about 710 g (25 oz; the mass of a gray squirrel, *Sciurus carolinensis*) in the largest of the sengis, the gray-faced sengi (*Rhynchocyon udzungwensis*). Sengis are restricted to central and eastern Africa, from about 15°N latitude southward. An exception is the North African elephant shrew (*Elephantulus rozeti*), which is found from Morocco to western Libya. Sengis are experts in gleaning invertebrates from the leaf litter. Their diet is diverse, including invertebrates such as beetles, crickets, spiders, ants, and termites; they also eat fruits and seeds. As with many species of rodents, a **cecum** is present.

In rufous sengis (*Elephantulus rufescens*), following the mating of the monogamous pair, one or two extremely **precocial** young are born after a **gestation period** of about 57 days (fig. 12.2B). Pups weigh 10 g (0.35 oz) and are able to walk and run within a day following birth (Rathbun, Beaman, and Maliniak 1981). They are born in and shelter in exposed spots at the bases of bushes, and the mother visits them to nurse only once a day until they are weaned. Could this rapid development be attributed to monogamy? Probably not—at least not directly. Cooperation between mates is less than perfect, and individuals spend very little time together. The limited collaboration between the sexes is very similar to that seen in some of the small African antelopes, klipspringers and dik-diks. Males and females do share the same territories; females will defend this area against other females, and males will evict intruding males. Galen Rathbun and colleagues found that captive mates and young seem to get along best when given plenty of space to accommodate their weak monogamous pair bond (Rathbun, Beaman, and Maliniak 1981).

What is the **adaptive** value of monogamy for sengis? It seems that the role of males as model housekeepers and mate keepers is crucial in rufous sengi ecology. The recent thinking is that monogamy in sengis is driven by male *mate guarding* (mate keeping), as described by Rathbun and Rathbun (2006). Sengis are primarily active at dawn and dusk, and like meadow voles (*Microtus*) of North American grasslands, sengis establish an elaborate system of runways employed for foraging and as avenues radiating to retreat areas. In eastern Africa, for instance, pairs of rufous sengis live in brushy and shrubby habitats and construct and maintain a complex system of trails through the brushy countryside. Rufous sengis are perpetually alert and very fastidious about trail maintenance—keeping roadways free of debris is crucial because sengis must run at full speed to avoid capture by keen-sighted hawks or other **diurnal** predators common in Kenya's dense woodlands. Roadwork is laborious and time consuming. Sengis spend a substantial number of daylight hours maintaining paths by sweeping matter aside with their forefeet. However, the allocation of time is not equal; males spend about 40% of their active daylight hours trail cleaning, compared with about 20% for their female mates (Rathbun 1979)—so males spend twice as long cleaning trails as do females. Keeping trails free of debris ultimately enhances reproductive success, for without well-manicured paths, foraging behavior in sengis (as in antelopes) would be greatly compromised and ineffective. Trail making and manicuring is best shown in rufous and four-toed sengis (*Petrodromus tetradactylus*). Most recent evidence indicates that the only sengis that build nests are the four species of *Rhynchocyon*. All other species find shelter in dense vegetation, rock cracks, or the shallow abandoned burrows of rodents. Some sengis may excavate their own shallow burrows (Rathbun and Redford 1981). Like small African antelopes, sengis spend their life exposed to the elements while relying on disruptive coloration to act as camouflage from the plethora of African predators.

The sengi freeways are dotted by less-traveled side roads marked with small, bare, oval "rest stops" where sengis may land for an infrequent break in their hectic work day. Sunbathing is commonly performed on resting sites along trails. Such sites are also used for scent marking by urination and defecation, and sengis will rub their sternal gland on the substrates to alert others to their presence. Rufous sengis are highly territorial and mark their boundaries by accumulating small piles of feces. Aggressive territorial encounters undoubtedly occur and are characterized by rear-foot drumming, fur fluffing, and strutting on tip-toe; in some cases, displays are followed by high-speed chases.

Sengis have long been a source of taxonomic controversy. The group was once included in the order "Insectivora" and, together with the tree-shrews (family Tupaiidae), in the suborder Menotyphla. Most biologists now believe that sengis belong in their own order, Macroscelidea (Schlitter 2005). Before the advent of molecular taxonomy, sengis were thought to share a common ancestry with rabbits and hares, but recent molecular methods applied to phylogenetic evidence indicate that sengis are actually part of a monophyletic African clade of mammals (supercohort Afrotheria) that includes elephants, sea cows, hyraxes, aardvarks, and golden moles and tenrecs (Springer et al. 2004).

Absentee Maternal Care

Treeshrews represent another excellent example of behavioral monogamy (Munshi-South, Emmons, and Bernard 2007). Treeshrews are in the order

Scandentia and represented by two families: Tupaiidae (treeshrews), containing 4 genera and 19 species, and Ptilocercidae (pen-tailed treeshrews), containing 1 genus and 1 species (Helgen 2005). Treeshrews are restricted to the Oriental faunal region, ranging from India, southern China, and the Philippines southward through Borneo and the Indonesian islands. Throughout their range, these mammals, resembling long-nosed tree squirrels (fig. 12.3), are found in forested habitats up to an elevation of 2,400 m (7,900 ft). It is important to note that the common name "treeshrew" is an unfortunate choice. Tupaiids are certainly not shrews and, although they are good climbers, they are not well adapted for **arboreal** life. Most treeshrews spend a great deal more time foraging on the ground than in trees. The larger species, such as the Mindanao treeshrew (*Urogale everetti*), confined to the island of Mindanao, and the large treeshrew (*Tupaia tana*) of Borneo and Sumatra spend most of their time on the ground. Treeshrews were first described in the late eighteenth century in an account by William Ellis, a surgeon accompanying the British explorer Captain James Cook on his voyage to the Malay Archipelago.

Several characteristics of treeshrews resemble those of primates, including their large braincase, **postorbital bar**, a permanent sac (scrotum) for the testes, and resemblance of the carotid and subclavian arteries to those of humans. However, most authorities consider treeshrews to be a distinct lineage separate from primates, as well as from "insectivores." Except for the pen-tailed treeshrew—which is exclusively **nocturnal** and equipped with large eyes (with a reflecting tapetum), large ears, long whiskers, and gray to black coloration adaptive for nocturnal activity—treeshrews are diurnal. Although chiefly **insectivorous**, treeshrews may consume fruits, seeds, and other plant matter to add extra calories or nutrients such as calcium to their high-protein diet (Emmons 1991). The larger treeshrews are known to consume lizards and small mammals in the wild.

Treeshrews are similar to sengis in that both have a maternal absentee system of neonatal care, but treeshrews are more extreme and unique. Research has confirmed that at least three species of treeshrews (pygmy, northern, and large) exhibit absentee maternal care coupled with social monogamy (Emmons 2000; Sargis 2004). Treeshrews build nests of dried leaves in hollow trees or nestled in root systems. Two nests are constructed, one for the parents and one for the young. The "natal" nest is made by the male—a departure from the usual mammalian scheme. Following a gestation period of 45 to 50 days, one to three young are born without fur and

FIGURE 12.3.   Treeshrews such as the large treeshrew (*Tupaia tana*) of Borneo and Sumatra (order Scandentia) exhibit a unique absentee maternal care coupled with social monogamy. Treeshrews are restricted to the Oriental faunal region, inhabiting forested habitats up to elevations of 2,400 m (7,900 ft). Several characteristics of treeshrews resemble those of primates. *Source:* Courtesy of Rod Williams / Bruce Coleman / Photoshot.

with closed ears and eyes. These **altricial** young are born in the natal nest. Another unconventional aspect of parental care in treeshrews pertains to the minimal amount of time the mother spends with her young. The mother visits her young only once every 48 hours to nurse them for 10 to 15 minutes; however, during this short visit, she is able to provide 5 to 15 g (0.17–0.52 oz) of milk to the rapidly growing young. This milk is high in protein (10%) to encourage rapid growth and has a very high fat concentration (25%) so that young can maintain a high body temperature, close to 37°C (98°F), despite the absence of the mother in the nest. The milk contains a small proportion of carbohydrate (2%) necessary to satisfy the immediate energy needs of the sedentary pups. About one month following birth, the pups disperse from their natal nest and join the parents in the parental nest. All in all, the mother spends only 1.5 hours with the infants during the nursing phase and shows no toilet care during this period. Young reach sexual maturity in about four months.

Lek Behavior

With their hammerlike head, guttural, blaring vocalizations, and unconventional courtship behavior, the hammer-headed fruit bats of Africa are unique among mammals (fig. 12.4). Hammer-headed bats (*Hypsignathus monstrosus*) are found along waterways in lowland tropical forests of Africa throughout the Congo basin to western Uganda; they are exclusively **frugivorous**, feeding on figs and fruits of several species of fever trees (*Anthocleista*). With a wing spread approaching 1 m (3.3 ft) and a body mass averaging about 420 g (15 oz), these are the largest continental African fruit bats. The genus exhibits the greatest sexual dimorphism of all bats, with males nearly twice the size of females (Bradbury 1977). Male hammer-headed fruit bats have a bizarre head shape associated with the competitive demands of mating, while females have a more conventional appearance.

Hammer-headed fruit bats are among the few mammals that exhibit a "lek" mating system, typified by an aggregation of displaying males to which females come for mating. The term is derived from the Swedish *leka*, meaning "to play." **Lek behavior** is common in birds such as sage grouse, prairie chickens, sharp-tailed grouse, and peacocks; it also occurs in hoofed mammals such as kob (*Kobus kob*), topi (*Damaliscus lunatus*), and fallow deer (*Dama dama*) as well as in some fishes and insects, including midges and ghost moths.

Male hammer-headed fruit bats assemble in leks at traditional locations every night during two annual dry seasons, June to August and December to February. Each night at sunset during the mating period, males leave their day roosts and fly directly to their traditional lek sites. Each aggregation includes 30 to 150 males, with a well-defined territory about 10 m (33 ft) in diameter. Singing usually occurs between 6:30 p.m. and 11 p.m. and between 3 a.m. and 5 a.m. The chorus emits a penetrating call described best as "guttural, explosive, and blaring" (Kingdon 1997b) and attracts females to the arena for the purpose of mating. At the lek site, males hang in the foliage at the edge of the canopy and emit their very loud, incessant croaking call, once to four times per second, accompanied by furious wing flapping. Fe-

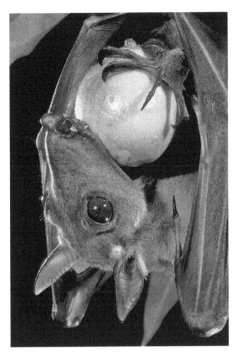

FIGURE 12.4.   The hammer-headed fruit bat (*Hypsignathus monstrosus*) is well known for its distinctive hammerlike head coupled with guttural, blaring vocalizations and somewhat unconventional courtship behavior, known as lek behavior. The hammer-headed fruit bat shown here is feeding on a rose apple. *H. monstrosus* is found along water courses in tropical forests of Africa. *Source:* Merlin D. Tuttle, Bat Conservation International.

males arrive and fly by the ranks, periodically hovering before a potential mate. Females mate with the most powerful singers that are positioned near the center of the choir.

The gregarious chorus is reminiscent of "a pond full of noisy American wood frogs, greatly magnified and transported to the tree tops" (Lang and Chapin 1917: 503). Kingdon (1997b: 117) refers to hammer-headed fruit bats as "flying loudspeakers," a well-deserved name given the males' greatly enlarged lips that function as megaphones to broadcast their position and intent (fig. 12.4). The noise generated by the choir can be heard within an 8 to 10 km (5–6 mile) radius of the lek. The male's vocal apparatus is highly specialized for production and resonance of noise. The unique and striking hammer-head appearance is due in part to the greatly modified lips and the elevated and enlarged nasal bones, accentuated by a large pouch that encloses the nose and extends over the cranium. The broad muzzle contains a pair of air sacs that open to the sides of the nose and can be inflated at will. The huge, tuba-shaped larynx contains large vocal cords and is positioned in the chest cavity, where it has displaced the heart, lungs, and diaphragm—the larynx is nearly one-half the length of the vertebral column!

Eusociality

Naked mole-rats (*Heterocephalus glaber*) are the only truly **eusocial** mammal. Other eusocial species are bees, ants, and wasps of the insect order Hymenoptera. Known to native Africans as "sand puppies," and unknown to western biologists until the mid-nineteenth century, African mole-rats (family Bathyergidae) are a strictly fossorial family of rodents, comprising 5 genera and 16 species, that inhabit the hot, sandy soils of Africa south of the Sahara Desert (Jarvis and Sherman 2002). The body of mole-rats is stocky and gopherlike in appearance, with a short tail and ears and minute eyes. Except for the naked mole-rat, which is practically hairless (fig. 12.5A), mole-rats have a thick, soft, and woolly or velvety pelage. Recent research indicates that mole-rats are indeed blind: their eyes are sightless; only the surface of the eye is used to detect air currents in the passageways. Like other fossorial species, mole-rats rely on touch receptors to navigate in their burrow systems. Many genera of mole-rats rely on long, touch-sensitive hairs throughout the body, augmented with highly sensitive senses of smell and hearing. The claws are of variable length depending on the mode of digging. The loose skin of mole-rats assists them in turning within their underground burrows—reportedly, they can almost perform a somersault as they turn and can rapidly move backward in tunnels. The skull is strong and robust. It is adapted to support their large **incisor** teeth. All species feed on **subterranean** tubers and bulbs of perennial plants, which are found by tunneling as far as 30 cm (almost 12 inches) below the surface. Mole-rats are no-nonsense tuber-eaters; in some habitats, tubers may weigh up to 50 kg (110 lb). Dune mole-rats establish very large caches of food in their burrows. All species burrow extensively, with tunnels of naked mole-rats (*H. glaber*) reaching more than 3 km (1.86 miles) in length (Brett 1986). Burrow systems are extensive, with numerous secondary branches and large chambers that serve for nesting, for **caching** food, and as toilet sites (Herbst and Bennett 2006).

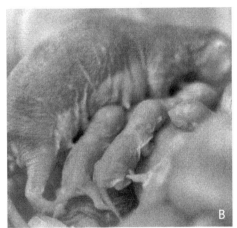

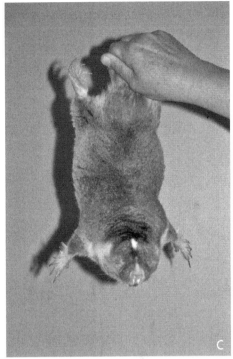

FIGURE 12.5. (A) Naked mole-rats (*Heterocephalus glaber*) are the only truly eusocial mammal. Known to native Africans as "sand puppies," mole-rats are strictly fossorial. (B) Naked mole-rat queen nursing her pups. (C) The Cape dune mole-rat (*Bathyergus suillus*) is the largest of the mole-rats, weighing up to 1.8 kg (3.9 lb). *Sources:* (A) Courtesy of John Visser / Bruce Coleman / Photoshot; (B) Stan Braude; (C) Hynek Burda.

Cape dune mole-rats (*Bathyergus suillus*) are the largest of the mole-rats, weighing up to 1.8 kg (3.9 lb) (fig. 12.5C). These heavy-bodied animals are endowed with some very impressive **procumbent** incisors; however, unlike the other mole-rats, their two lower incisors cannot be moved independently, as the lower jaws are ossified at the symphysis (articulation) of the mandible. Because of their large size and "flawed" mandibular morphology, dune mole-rats do not dig with their incisors but rather use forepaws equipped with long claws. These two traits seem to restrict them to habitats with sandy, less compact soils. All other mole-rats rely on their powerful procumbent incisors for both digging and defense. As in other species of fossorial mammals, soil is prevented from entering the mouth during digging by closure of the lips behind their incisors.

Several species of bathyergids are solitary, including the Namaqua dune mole-rat (*Bathyergus janetta*) and the silvery mole-rat (*Heliophobius argenteocinereus*). Other species, such as the Togo mole-rat (*Cryptomys zechi*) of east central Ghana and adjacent Togo, are found in pairs or small groups (Yeboah and Dakwa 2002), while the Damara mole-rats (*C. damarensis*) average about 18 individuals in a colony. However, as we will see, the largest colonies are reported for naked mole-rats (*H. glaber*) residing in Kenya: numbers in a burrow average about 75 to 80 and can reach as many as 300 individuals (Lacey and Sherman 1997).

Because mole-rats reside in a relative stable and protected subterranean world, predators are not a serious problem. As a consequence, litter size is small, ranging from two to five young; however, there are exceptions. Following a gestation period of 44 days, cape mole-rats produce 3 to 10 young. They have two litters per season. Dune mole-rats have a slightly longer gestation period of 52 days and bear 1 to 4 young twice per season. Following a gestation period of about 70 days, a litter of up to 28 blind (and, of course, naked) young are born. Females are reported to have 15 **mammae** (Sherman, Braude, and Jarvis 1999). Despite their very impressive reproductive performance, mole-rats must settle for second place. First place goes to the tail-less tenrec (*Tenrec ecaudatus*) of Africa, described earlier in the chapter, which is reported to produce up to 32 young in one litter.

Naked mole-rats are unique in many ways. They live in colonies of up to 300 animals, all of the same family, with a strict division of labor overseen by a single breeding female. Naked mole-rats also have the longest known lifespan for any **terrestrial** animal weighing less than 100 g (3.5 oz): they are reported to live as long as 16 years or more in captivity. They are essentially naked, possess only tactile whiskers, and lack sweat glands and (unlike most mammals,) subcutaneous fat. The latter features contribute to their substandard thermoregulatory constitution and qualify them as physiologically **poikilothermic**—cold-blooded, with a body temperature that fluctuates with that of the environment (Buffenstein and Yahav 1991).

About the size of a meadow vole (*M. pennsylvanicus*), naked mole-rats are the smallest of the mole-rats, with head and body length of 9 to 12 cm (3.5–4.7 inches) and an average mass of about 35 g (1.2 oz). They inhabit arid regions of Kenya, Ethiopia, and Somalia in northeastern Africa. Their lack of fur, coupled with a reduced **metabolic rate**, greatly facilitates dissipation of body heat while underground. *Heterocephalus* maintains a low body temperature of about 32°C (89.6°F) in its underground world. Regardless of the environmental temperatures aboveground, the burrows usually range between 30°C and 32°C (86°F–89.6°F), with a relative humidity of about 90%. Activity of naked mole-rats is usually timed for early morning and late afternoon to avoid extreme temperatures.

At least two species of mole-rats, the naked mole-rat (*H. glaber*) and the Damara mole-rat (*C. damarensis*), have colony structures similar to those of social insects (Jarvis 1981). Damara mole-rat colonies average 16 individuals with a single breeding female, whereas naked mole-rat colonies may have up to 300 individuals. Jennifer Jarvis first demonstrated mammalian eusociality in the naked mole-rat (Jarvis 1981; Sherman, Jarvis, and Alexander 1991). She captured 40 members of a colony from their burrow system in Kenya and studied them for six years in an artificial burrow system in the laboratory. Only one female in the colony at any one time (the "queen") had young; mother and young were fed but not nursed by male and female adults of the worker caste; members of this caste were not seen to breed. Another caste of nonworkers assisted in keeping the young warm; males of this caste bred with the reproducing female. Mole-rats cooperate in tunnel digging, predator defense, and reproduction. Individuals vary greatly in body size, depending on their function in the colony—smaller workers versus larger nonworkers. Some individuals remain workers throughout their lives, while others grow larger and become defenders of the colony. These larger and older individuals will represent the "pool" of new reproductive males when the prime breeding male dies. The one reproductively active female is the largest individual and mates with one of very few reproductively active males in the colony. Other adults do not breed but take part in foraging, caring for young, and providing other cooperative functions. Like rabbits, shrews, some rodents, and mountain beavers, naked mole-rats practice **coprophagy** (feeding on feces). This practice may be essential for the young to acquire the bacteria they need in their alimentary canal to digest plant food. Nonbreeders are not sterile and can rapidly become sexually mature (within 7–10 days) and disperse to establish new colonies or replace breeders that die. Stan Braude of Washington University in St. Louis found that naked mole-rats will disperse more than 2 km (1.2 miles) from their natal colony (Braude 2000; O'Riain and Braude 2001), and this may contribute to the normal pattern of outbreeding found in the wild (Hess 2004).

Researchers have found that when the mole-rat queen dies, the next largest and strongest female in the colony starts to grow, her mammae enlarge, and she assumes breeding duties. What prevented this female from establishing a nest chamber and producing young of her own while the old queen was alive? Researchers believe that the "queen" of the colony secretes a unique chemical inhibiting reproductive activity in colony members. Pheromones do not seem to be the prime means of birth control, as they are in social insects, yet the entire colony is usually influenced by the reproductive state of the breeding female. The precise mechanism is unknown and may be related to stress through behavioral interaction between subordinate and dominant individuals (Faulkes and Abbott 1993). The entire colony seems to be influenced by the reproductive behavior of the queen. Just before she produces a litter, all colony members (both males and females) develop mammary glands, with some females actually entering breeding condition. These changes seem to be important in priming the colony members as surrogate parents.

Digging by naked mole-rats is accomplished primarily with their impressive forward-pointing incisor teeth; the feet are employed to kick and push the loosened soil. Digging is a cooperative effort with "assigned tasks." Families line up, nose-to-tail, forming "digging chains" analogous to conveyor belts. The lead excavator uses its teeth to break through and loosen soil, kicking it backward to several sweepers, who then whisk the soil to the rear. At the back of the line, a mole-rat violently kicks the soil to the surface, forming a distinctive volcano-shaped mole hill. During the digging process, animals in the digging chain advance in position by straddling the excavator as it moves to the rear.

Semelparity: Breeding and Sudden Death

Small mammals allocate a great deal of energy for maintenance, growth, foraging activity, and reproduction. Life history traits such as size at birth, litter size, age at maturity, and degree of parental care directly influence fecundity and survivorship.

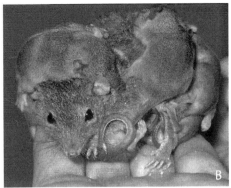

FIGURE 12.6. Of all the marsupial carnivores of the family Dasyuridae, the genus *Antechinus* is the most numerous and widespread in Australia. (A) The brown antechinus (*Antechinus stuartii*) exhibits a rare reproductive behavior: it reproduces only once in its lifetime, and all females come into estrus at the same time just once a year. After mating, all the males die. (B) Brown antechinus female and young. *Source:* Fritz Geiser.

Natural selection recognizes only one form of currency: successful offspring. Reproductive success is paramount. The reproductive strategy employed by some organisms, such an annual plants, insects, and Pacific salmon that reproduce only once in their lifetimes, is called **semelparity** (from the Latin *semel,* "once," and *pario,* "to beget").

Semelparity is contrasted with **iteroparity** (from the Latin *itero,* "to repeat"), the production of offspring in successive bouts throughout an organism's lifetime. Iteroparity is practiced by most species of mammals. The semelparous, or "big bang," breeders among mammals include American didelphids (*Marmosa, Marmosops,* and *Monodelphis*) and three groups of Australian dasyurids: the antechinuses, phascogales, and northern quolls (Cockburn 1997; Oakwood, Bradley, and Cockburn 2001; Leiner, Setz, and Silva 2008). Among these semelparous mammals, we know the most about the brown antechinus (*Antechinus stuartii*) (fig. 12.6).

The brown antechinus is a small insectivorous dasyurid **marsupial** that inhabits the forests of eastern Australia. These small dasyurids range in mass from about 17 to 71 g (0.59–2.5 oz); the males typically are 20% to 100% heavier than the females. Day length acts as the trigger for the onset of reproductive activity, which begins in August in southern Australia and September in southern Queensland. Mating takes place in the same week each year for any population. All animals mate in the first breeding season when males have reached 11 to 12 months of age. Males become active and aggressive during the breeding season. During this short, two-week season, as many as 20 males may aggregate in a tree cavity where they are visited by the females. Copulation is intense and prolonged, generally occurring for about six hours, with a few of the males mating many times. At the end of the two weeks, not a single male is left alive.

What accounts for this sudden death in *Antechinus?* Death results from stress associated with the social demands of the mating season, during which time the males stop feeding and live on their reserves while seeking every opportunity to mate. It is believed that an increased physiological stress results from this heightened activity, aggression, and competition for mates. Increasing levels of testosterone and other androgens evidently depress plasma-corticosteroid-binding globulin, resulting in a rise in corticosteroid concentration. The trauma associated with stress and endocrine changes causes suppression of the immune system, major ulceration of the gastric mucosa, increased vulnerability to parasites of the blood and intestines, and bacterial infections of the liver.

Pregnant females live on, and following a gestation of about four weeks, produce a litter of 6 to 10 tiny young weighing some 16 mg (0.0006 oz) each. Females have a shallow pouch with 6 to 10 teats. The young suckle for about 5 weeks, but weaning does not take place until 14 weeks of age. Prior to weaning, young are dragged along the ground rather awkwardly, but this adventure does not seem to adversely influence their survival. Following weaning, the young are left in a spherical nest of plant material hidden in a hollow log or tree trunk. Females may breed in a second season; however, the success of reproduction in subsequent years is not favorable. Excellent discussions of the antechinuses and other "big bang" breeders are provided by McAllan (2003), Tyndale-Biscoe (2005), Foster et al. (2008), and Van Dyck and Strahan (2008).

CHAPTER THIRTEEN

Population Cycles:
Lemmings and Snowshoe Hares

Voles and lemmings (subfamily Arvicolinae) are touted as one of the most prolific groups of small mammals: a single female meadow vole may produce up to 72 young in one breeding season. A population of meadow voles in Maryland was reported to have increased from a density of less than 3/ha (1 hectare = 2.47 acres) to nearly 400/ha (a 133-fold increase) over a four-year period (Jett and Nichols 1987). To my knowledge, the record for vole density goes to an Oregon population of montane voles (*Microtus montanus*) that reportedly reached 9,900/ha (4,000/acre) (Spencer 1958). Arvicoline populations have a long history of population "irruptions," and lemmings such as the Siberian brown lemming (*Lemmus sibiricus* [= *obensis*]) on the Yamal Penninsula in northwest Russia show remarkable productivity, with a density of up to 2,000/ha (810/acre) (Karaseva et al. 1971).

It is safe to say that lemmings reach the highest population densities of all tundra rodents. According to legend, at the time of a population explosion, lemmings reduce their numbers by the thousands by flinging themselves into the sea. This is, indeed, just legend; the story is not true. Lemmings do not undergo premeditated suicidal marches into the sea! During years of high density, lemmings will, in fact, disperse. When they reach a body of water such as a river, lake, or arm of the sea, they may attempt to cross to the other side, and although lemmings are excellent swimmers, many will drown. The myth of mass suicides has a long history popularized by several factors, and Walt Disney contributed little to our understanding of lemming biology. The 1958 documentary *White Wilderness* showed footage of lemmings migrating and dashing headlong over a ledge into the sea; the documentary was filmed in Alberta, Canada (a landlocked Canadian province where no lemmings are found). The filmmakers faked the entire sequence, using imported lemmings and a snow-covered turntable on which a few dozen lemmings were forced to run—some were literally thrown into the "sea" to show the alleged suicides. Unfortunately, things have not improved much. You can now participate in a popular interactive computer game, *Lemmings,* in which the player attempts to save the mindlessly marching lemmings from walking to their deaths.

Myths and misconceptions about lemmings go back many centuries (Marsden 1964; Finerty 1980). The first known depiction of lemmings dates from the middle ages. An illustration by Archbishop Olaus Magnus of Uppsala depicts lemmings falling from the clouds, and a scene on his great map of 1539 shows two stoats (one in a trap) each with a lemming in its jaw. (The first mention of migration in lemmings also dates from the sixteenth century.) The writer Jakob Kruger claimed that lemmings indeed fell from the sky, in his work "Strange Apparition of Mice Which Fell in Norway from the Air to the Earth on to Houses" (translated from the German). During the autumn of 1579, lemmings (called "lemmen" by the local inhabitants) reportedly fell, on frequent occasions, from the air to the ground, into the water, and onto houses in Bergen. Many lemmings were collected by merchants to "prove" that the event was not a fable but a true event. Further "proof" that lemmings fell from the sky was obtained by Peer Clausson Friis in 1599. The notion of lemmings falling from the sky was not limited to the Old World, as it is also present in Eskimo myth (Marsden 1964: appendix A). The history (as opposed to myth) of lemming migrations and the researchers who have studied lemming cycles is well documented by Marsden (1964), Krebs and Myers (1974), Finerty (1980), and Chitty (1996).

Beginning in the early 1920s, the British ecologist Charles Elton (1924) initiated a rigorous scientific study of the existence of regular, long-term fluctuations in populations of small mammals. Since that time, mammalian ecologists have been mystified by "microtine cycles" ("microtine" is derived from "Microtinae," an earlier name for the subfamily that includes voles and lemmings), yet no clear consensus has been reached on the cause(s) of these cycles (Krebs 1996). In 1987, Bill Lidicker presented the Plenary Address at the annual meeting of the American Society of Mammalogists; he was the recipient of the society's prestigious C. Hart Merriam Award. In the light of the title of his presentation, "Solving the Enigma of Microtine 'Cycles,'" I believe it is safe to say that vole population cycles are indeed one of the best-known "unsolved mysteries" in mammalian ecology (Lidicker 1988).

Population cycles range from 3 to 4 years in smaller species such as voles and lemmings (*Lemmus* and *Microtus*) up to 10 years or more in larger species such as lynx (*Lynx canadensis*) and snowshoe hares (*Lepus americanus*) (fig. 13.1B). The vole population cycle typically spans 3 to 4 years; the Nearctic brown lemming (*Lemmus trimucronatus*), inhabitant of the tundra near Barrow, Alaska, is one such cyclic species (fig. 13.1B, top). The lemming cycle consists of low, increase, peak, and decline phases. During one of the increases, breeding starts in the autumn, and by spring, large numbers of lemmings are present, many of which fall prey to snowy owls (*Nyctea scandiaca*), arctic foxes (*Alopex lagopus*), weasels (*Mustela*), and other predators. During the summer, the population crashes to a low level, where it remains for 1 to 3 years.

Two general sets of factors—extrinsic factors such as changes in food availability (Pitelka 1958), parasites, disease, predators, and weather, and intrinsic factors such as physiological stress from crowding and changes in gene frequency (often called the Chitty hypothesis)—have been proposed to explain rodent cycles (Krebs and Myers 1974; Lidicker 1988; Stenseth and Ims 1993; Krebs 1996). Many researchers have focused on one or two factors in an effort to explain cycles. For example, in eastern North America, the early work of W. J. Hamilton (1937) demonstrated that meadow voles (*Microtus pennsylvanicus*) exhibited a four-year cycle. Vole numbers fluctuated from a low of 15/acre to a peak of 250/acre. The increase was attributable to accelerated breeding rate, increased numbers of young per litter, and the lengthening reproductive season. The decrease in numbers was due to factors such as climate, disease, and predators. The primary causative agent for the decline was murine epizootics (disease outbreaks affecting many murine animals). In Scandinavia, Ilkka Hanski and coworkers argued that small, **specialist** predators such as weasels, the major predators at high latitudes, increase the amplitude of the cycles, and that larger, **generalist** predators, which are more important at lower latitudes, tend to stabilize vole populations (Hanski, Hansson, and Henttonen 1991). In the high arctic tundra of northeast Greenland, the collared lemming (*Dicrostonyx groenlandicus*) is the main source of food for no less than four vertebrate predators, making this one of the world's simplest predator-prey communities (Gilg 2002; Gilg et al. 2006). Lemmings follow a four-year cycle, with peak population densities more than 100 times trough densities.

Longer-term cycles, from 8 to 12 years, are known in populations of snowshoe hares (*L. americanus*) and Canada lynx (*L. canadensis*) in boreal

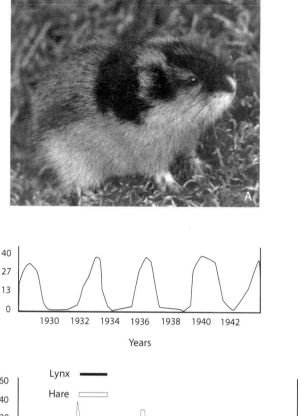

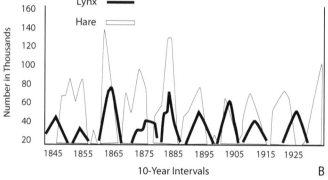

FIGURE 13.1. Population cycles. (A) The Norway lemming (*Lemmus lemmus*) of Finno-Scandia is a cyclic mammal. (B) Investigators have estimated lemming densities for a close relative, the Nearctic brown lemming (*Lemnus trimucronatus*) in the coastal tundra at Barrow, Alaska, for a 20-year period (top graph). Population peaks are commonly separated by an interval of 3 to 4 years. The bottom graph shows long-term cycles, from 8 to 12 years, in populations of snowshoe hares (*Lepus americanus*) and Canada lynx (*Lynx canadensis*) in boreal forests and tundra of North America. *Sources:* (A) Asko Kaikusalo; (B) Carie Nixon, Illinois Natural History Survey.

forests and tundra of North America (fig. 13.1B, bottom). Some of the first data on lynx-hare cycles were presented by E. T. Seton (1911) in his book *The Arctic Prairies,* based on the numbers of furs brought in by trappers to the Hudson's Bay Company (MacLulich 1937). For example, in 1865, the Hudson's Bay Company was inundated with snowshoe hare pelts, but by 1870 very few pelts were being delivered by trappers. In 1875, hare pelts poured in once again in huge numbers, and this event was repeated in 1885, 1895, 1905, and 1915. The company records of fur returns showed peaks followed by precipitous declines. The returns had nothing to do with fur prices, but rather reflected the biology of the species. Each cycle is about 10 years in length, with the lynx and hare populations highly synchronized— the peaks in lynx numbers following those of the hare by a year or two. How do we explain such predictable patterns? One hypothesis for the cause of these cycles examines time delays in the interaction of three trophic (feeding) levels: the winter food supply for hares, the number of hares, and the number of predators (primarily lynx and great horned owls, *Bubo virginianus*). A second hypothesis includes only two trophic levels: hares and their predators.

Mathematical models and laboratory experiments show that cycles can occur in the absence of any regular environmental fluctuations. Keith (1983) concluded that food shortage during the winter initiates the decline in hare density, with predators playing a secondary role. Lynx density seems to depend on hare density, so the lynx is food-limited, but whether hares are limited only by predators or by both food and predators is unclear (Keith 1987). In an experiment designed to test the role of food and predators in hare density, researchers manipulated supplemental food and mammalian predator abundance on 1 km² (0.39 square mile) plots in the Yukon (Krebs et al. 1995). Mammalian predators were excluded from some plots by means of electric fences. Hare density during the cyclic peak and decline doubled when predators were excluded and tripled when food was added. When predator exclusion was combined with addition of food, hare density increased 11-fold. Thus, both food supply and predation seem important in population cycles of hares, and they seem to interact in a synergistic, or nonadditive, way. These results tend to support the three-trophic-level hypothesis.

Whatever the immediate causes of snowshoe hare cycles, the observation that cycles frequently are synchronized in time and space also requires an explanation. Could some continentwide synchronizer be affecting climate and thus altering the hare's food supply over a wide region? One candidate cause suggested early on was peaks in sunspot activity, which have about an 11-year mean. The suspected role of sunspots was later discounted because the mean of the hare cycles was closer to 10 years (Keith 1963). However, augmented by more recent analyses, this theory has been resurrected (Sinclair et al. 1993; Sinclair and Gosline 1997).

In his discussion of population cycles, Steve Vessey concludes: "After decades of work by mammalian ecologists from many countries, including some studies of voles spanning more than 25 years, researchers know much about what are *not* the causes of cycles" (in Feldhamer et al. 2007: 477). As we track historical records, it is clear that lemming "outbreaks" have been a feature of northern latitudes for the past millennium. It is noteworthy that in some areas these outbreaks have either stopped or occur less frequently (Coulson and Malo 2008; Ims, Henden, and Killengreen 2008). The cause of this change is unclear, but in parts of northern Europe, some factor, possibly climate change, is preventing the regular occurrence of population cycles of arvicoline rodents (Kausrud et al. 2008).

Appendix

Useful Web Sites

American Society of Mammalogists
www.mammalsociety.org

Bat Conservation International
www.batcon.org

Natural History Museums and Collections
www.lib.washington.edu/sla/natmus.html

The Shrew (-ist's) Site
http://members.chello.at/natura/shrew/index.html

University of Michigan Museum of Zoology
http://animaldiversity.ummz.umich.edu

Acta Chiropterologica
www.miiz.waw.pl/periodicals/acta-chiropterologica/

Acta Theriologica
http://acta.zbs.bialowieza.pl/

Mammalia
www.degruyter.de/journals/mammalia/detail.cfm

Mammalian Species
www.acgpublishing.com/dir_Journals/MammalianSpecies2.asp

Glossary

....

abomasum. The fourth and last chamber of the stomach of a ruminant; often called the "true stomach" in other mammals. Most of the absorption occurs in this chamber.

adaptation (adaptive). A peculiarity of structure, physiology, or behavior of an organism that aids in fitting the organism to its particular ecological niche or habitat.

adaptive hypothermia. A group of energy-conserving responses of mammals and birds characterized by the temporary abandonment of mechanisms that maintain a constant, normal body temperature (homeothermy).

Allen's rule. The ecogeographic "rule" stating that in endothermic animals, the extremities are shorter in animals living in colder climates than in the same species found in warmer climates.

altricial. Describing young that, at birth, are blind, naked, immobile, and entirely dependent on parental care; as in most mammals and many birds. Contrast **precocial.**

apocrine sweat glands. Type of sweat gland found on the palms of the hands and bottoms of the feet in mammals; highly coiled structures located near hair follicles. Contrast **eccrine sweat glands.**

arboreal. Living in or frequenting trees.

auditory bulla (pl. **bullae**). Bony case enclosing the sensory apparatus of the ear.

baculum. Stiffening rod of bone in the penis of many mammalian groups. Also called the *os penis.*

Bergmann's rule. The ecogeographic "rule" stating that races of a warm-blooded species living in warmer climates are smaller than races found in cooler climates.

bicornuate uterus. Type of uterus in placental (eutherian) mammals, with a single cervix and the two uterine horns fused for part of their length; as in "insectivores," most bats, primitive primates, pangolins, some carnivores, elephants, manatees, dugongs, and most ungulates.

brachyodont. Having low-crowned teeth—teeth not as high as they are wide or long. Contrast **hypsodont.**

brown adipose tissue. A special type of fat packed with mitochondria; the site of nonshivering thermogenesis in placental (eutherian) and some marsupial (metatherian) mammals. The tissue is essential in promoting the increased metabolism necessary for arousal from hibernation. Also called *brown fat.*

bunodont. Having low-crowned teeth with rounded, blunt cusps; used primarily for crushing.

caching. The handling of food for the purpose of conserving it for future use. Also called *food hoarding.*

calcar. In bats, a spurlike, cartilaginous process that extends medially from the ankle and helps support the uropatagium.

canine. Describing the four pointed teeth in the jaws of many mammals. One canine tooth is located on each side of the upper and lower jaw, just behind an incisor.

carnassial. Describing bladelike, shearing cheekteeth found in most carnivores; the last upper premolar and the first lower molar in extant mammals. These teeth are most highly developed in felids and canids.

carnivorous (carnivory). Having a diet that consists mainly of animal material.

cavernous sinus. A network of small vessels immersed in cool venous blood, located in the floor of the cranial cavity, that is important in heat exchange in certain carnivores and artiodactyls.

cecum (pl. ceca). The blind sac or pouch in which the large intestine of many mammals begins; the appendix in humans.

cellulolytic. Cellulose-splitting. Cellulose is broken down by enzymes such as cellulase.

circumpolar. Located in one of Earth's polar regions.

cloaca. A chamber at the terminal part of the gut into which the reproductive, digestive, and urinary systems empty and from which waste products leave the body.

commensal. Participating in commensalism, a kind of symbiosis in which one species benefits while a second species (the host) is neither harmed nor benefited.

convergence. The evolutionary tendency of organisms that are not closely related to develop similar adaptations when subjected to similar environmental conditions and constraints.

coprophagy. Feeding on feces, occurring in shrews, lagomorphs, and rodents. Also called *refection; reingestion.*

countercurrent heat exchange. An arrangement of blood vessels that allows peripheral cooling, particularly of appendages, and at the same time maintains an adequate blood supply without excessive heat loss.

crepuscular. Active during periods of twilight, whether dawn or dusk. Contrast **diurnal; nocturnal.**

cursorial. Adapted for running locomotion.

cusp. In mammalian teeth, a projection on the biting surface of a molar.

daily torpor. A form of physiological heterothermy typified by bouts of torpor limited to a duration of less than 24 hours.

Dehnel's phenomenon. The observation that reduction in the body size of the common shrew (*Sorex araneus*) in autumn is accompanied by shrinkage in the size of the skull.

delayed implantation. Suspension of embryonic development after the first few cell divisions and before the embryo embeds (implants) in the wall of the uterus.

diastema. A gap between adjacent teeth, such as between incisors and cheekteeth, in rodents, lagomorphs, artiodactyls, and perissodactyls.

digitigrade. Walking on the toes with the heels not touching the ground, as in cats and dogs. Contrast **plantigrade**.

dilambdodont. Having tooth cusps and associated ridges arranged in a W-shaped pattern.

diurnal. Active during the daylight hours. Contrast **crepuscular; nocturnal**.

Doppler shift. The apparent change in sound or light frequency caused by movement of the source or the receiver of sound or light waves.

duplex uterus. Type of uterus in which the right and left parts are completely unfused and each has a distinct cervix; found in lagomorphs, rodents, aardvarks, and hyraxes. Contrast **simplex uterus**.

eccrine sweat glands. Type of sweat gland with separate ducts that lead to the body surface, through which water is forced outward. Found throughout the body, they are important as a means of evaporative cooling in mammals. Contrast **apocrine sweat glands**.

echolocation. A process by which an animal orients itself, or identifies the location, character, and perhaps movement of objects, by emitting high-frequency sounds and interpreting the reflected sound waves. Used by many bats and shrews in navigating and in locating prey.

ectothermic (ectothermy). Having a body temperature that approximates that of the environment. Also called *poikilothermic; cold-blooded*. Contrast **endothermic (endothermy)**.

embryonic diapause. A period in development of the embryo at the stage of the blastocyst (70- to 100-cell stage).

endothermic (endothermy). Maintaining a relatively constant body temperature by means of heat produced inside (*endo*) the body; emphasis is on the mechanism of body temperature regulation. Also called *homeothermic; warm-blooded*.

epitheliochorial placentation. Arrangement of the chorioallantoic placenta typified by six tissue layers, with the villi resting in pockets in the endometrium; the least modified placental condition.

estivation. A form of physiological heterothermy in which the body temperature and rate of metabolism are reduced; an adaptation for reducing energy expenditure in periods of heat or shortage of water.

estrus. The period during which a female mammal is sexually receptive to a male. Also called *heat.*

eusocial (eusociality). Having a social system involving reproductive division of labor and cooperative rearing of young by members of previous generations.

euthermy. The condition in which mammals periodically increase their body temperature from near 0°C to approximately 36°C (from 32°F to 96.8°F).

evaporative cooling. Cooling due to absorption of heat when water changes state from liquid to vapor. Heat is absorbed from the surface at which the change of state occurs and is carried away with the water vapor produced.

folivorous (folivory). Having a diet composed primarily of plant leaves.

foregut fermentation. The digestive process in mammals that have a complex, multi-chambered stomach with cellulose-digesting microorganisms that enable the host animal to derive nutrients from highly fibrous foods. Also called *rumination.* Contrast **hindgut fermentation.**

fossorial. Adapted for digging and burrowing and spending much time beneath the surface of the ground; as in moles and pocket gophers.

frugivorous (frugivory). Consuming a diet of fruit.

generalist. Having a lifestyle that does not involve highly specialized strategies; for example, feeding on a variety of foods that may require different foraging techniques. Contrast **specialist.**

gestation period. The period of development of the embryo in the uterus.

Gloger's rule The ecogeographic "rule" stating that races in warm and humid areas are more heavily pigmented than those in cool and dry areas.

granivorous (granivory). Consuming seeds.

guard hairs. The stiffer, longer, outer hairs of the pelage, overlying the underfur.

gummivorous (gummivory). Consuming plant resins, sap, or gums.

heat load. The sum of environmental and metabolic heat gain.

hemochorial placentation. The arrangement of the chorioallantoic placenta that lacks maternal epithelium; villi are in direct contact with the maternal blood supply; found in advanced primates (including humans), "insectivores," bats, and some rodents.

herbivorous (herbivory). Consuming any type of plant material.

heterothermic (heterothermy). Exhibiting, at some times, a well-regulated body temperature (homeothermic) and, at other times, a body temperature close to that of the environment (ectothermic).

hibernation. A form of physiological heterothermy that occurs during the winter season and is characterized by a series of prolonged bouts of torpor in which body temperature is usually very close to environmental temperature.

hindgut fermentation. A digestive system in which food is completely digested in the stomach and passes to the large intestine and cecum, where microorganisms ferment the ingested cellulose. Contrast **foregut fermentation**.

home range. The area that an animal traverses during its normal activities of food gathering, mating, and caring for its young

homeothermic (homeothermy). Maintaining a constant body temperature.

hypothermia. A condition in which the temperature of the body is subnormal.

hypsodont. Having high-crowned teeth—teeth higher than they are wide or long. Contrast **brachyodont**.

hystricomorph. Rodent in which the infraorbital foramen is greatly enlarged; porcupine- or cavy-shaped.

implantation. The process in which the blastocyst (embryo) embeds in the lining of the uterus.

incisors. Usually unicuspid teeth anterior to the canines that are used for cutting or gnawing.

incrassated. Thickened or swollen.

incus. The second of the three bones of the middle ear (ossicles) in mammals.

infraorbital foramen. An opening in the skull below the orbit, or eye socket.

insectivorous (insectivory). Feeding on insects.

insensible water loss. The mechanism by which water is lost by diffusion through the skin and from the surfaces of the respiratory tract; also called *transpirational water loss*.

iteroparity (iteroparous). Production of offspring in successive bouts throughout the lifespan; occurs in most species of mammals. Contrast **semelparity (semelparous)**.

K-selection. Selection favoring slow rates of reproduction and growth, characteristics adapted to stable, predictable habitats. Contrast **r-selection**.

lactation. Production of milk by mammary glands (mammae).

lek. An area used, usually consistently, for communal courtship displays.

lophodont. An occlusal pattern in which the cusps of cheekteeth form a series of continuous, transverse ridges, or lophs; as in elephants.

lower critical temperature. The temperature at which an animal must increase its metabolic rate to balance heat loss.

malleus. The first of the three bones of the middle ear (ossicles) in mammals. The "hammer" connects the tympanic membrane (eardrum) and the incus.

mamma (pl. mammae). One of the two or more milk-producing glands unique to mammals, developing in both sexes but rudimentary in males. Also called *mammary gland*.

marsupial. Describing a mammal that has a marsupium, an external pouch formed by folds of skin in the abdominal wall, enclosing mammary glands (mammae) and serving as an incubation chamber for precocial young. Besides "marsupial" mammals, a marsupium is also found in echidnas.

masseter. One of three main masticatory muscles of mammals that functions to close the mouth by raising the mandible. Pronounced in herbivorous mammals, the masseter aids with the horizontal movement of the jaw.

mass-specific metabolic rate. The rate of energy necessary per gram of body mass; refers to energy demands in an animal's tissues.

mesic. Of a soil or habitat, moderately moist. Contrast **xeric.**

metabolic rate. Energy expenditure, measured in kilojoules per day.

metabolic water. Water produced by aerobic breakdown (catabolism) of food. Also called *oxidation water.*

metabolism. The sum of the chemical processes in an organism by which energy is provided for vital processes and activities.

midden. (1) A refuse heap; specifically, a pile of guano, food, or other material, such as the piles of plant material, bones, and so on, deposited by woodrats. (2) The stored cones or other food of squirrels.

molar. Any of the hindmost teeth of the jaw in most mammals, typically robust and adapted for crushing, which are situated behind the premolars; not developmentally preceded by milk teeth. Contrast **premolar.**

monestrous. Having a single estrous cycle (period of heat in the female) per year. Contrast **polyestrous.**

monogamy. Mating system in which animals have a single mate for life, or at least for one breeding season. Contrast **polyandry; polygamy; polygyny.**

mycophagous (mycophagy). Consuming a diet of fungi.

mycorrhiza. The mutualistic association of the mycelium of a fungus with the roots of a tree or other seed plant.

myomorph. Rodents in which the infraorbital foramen is small to moderate in size; mouse-shaped.

myrmecophagous (myrmecophagy). Feeding primarily on colonial insects such as ants and termites. Many mammalian families are primarily or secondarily myrmecophagous.

nectarivorous (nectarivory). Consuming a diet of nectar; found in about six genera of bats and marsupial honey possums.

nocturnal. Active during the night. Contrast **crepuscular; diurnal.**

nonshivering thermogenesis. Means of heat production in mammals that does not involve muscle contraction. The source of nonshivering thermogenesis is brown adipose tissue.

norepinephrine. A catecholamine neurohormone of mammals that stimulates production of heat by brown adipose tissue. Also called *noradrenaline.*

occlusal. Describing surfaces of upper and lower teeth that come in contact during chewing. Occlusal surfaces have one or more cusps (unicuspid or multicuspid).

omasum. The muscular third chamber of the stomach of a ruminant.

omnivorous. Feeding on both animal and vegetable foods.

panting. A method of cooling characterized by very rapid, shallow breathing that increases evaporation of water from the upper respiratory tract; occurs in canids and small ungulates.

patagium. The gliding membrane in, for example, marsupial gliders, flying squirrels, bats, and colugos that typically stretches down the side of the body between the forelimb and hind limb and occasionally includes the tail.

pectinate. Comblike, describing a structure with several prongs or projections in a row.

pelage. The entire furred coat of a mammal, whether of the living animal or of a dead animal prior to skinning; the entire furred coat after its removal is referred to as the *pelt*.

pentadactyl. Having five digits. The hands and feet of humans are pentadactyl, as are those of "insectivores."

piloerection. Fluffing of the fur.

pinnae. External ears that surround the auditory canal and channel sound waves to the tympanic membranes (eardrums); not found in many marine and fossorial mammals.

piscivorous (piscivory). Consuming a diet composed primarily of fish; characterizes bulldog bats (*Noctilio*).

placental. Having a placenta, the vascular structure formed by the union of the uterine tissue of the mother and the membranes of the developing embryo; provides for the transfer of gases, nutrients, and waste materials between mother and fetus. Only placental (eutherian) mammals have a well-developed placenta. Marsupials have a rudimentary or no placenta; monotremes lay eggs.

plantigrade Walking on the soles of the feet with the heel touching the ground; as in bears and humans. Contrast digitigrade.

poikilothermic (poikilothermy). Ectothermic, cold-blooded; having a body temperature that varies with that of the environment.

polyandry. Mating system in which a female has more than one mate. Contrast polygyny.

polyestrous. Having more than one estrous cycle (period of heat) per year. Contrast monestrous.

polygamy. Mating system in which individuals have more than one mate; a general term in which gender is not specified. Contrast monogamy.

polygyny. Mating system in which one male mates with several females. Contrast polyandry.

postorbital bar. A bony rod separating the orbit from the temporal fossa. Formed by a union of the postorbital process of the frontal bone with that of the zygomatic arch.

precocial. Describing young born in a relatively well-developed condition—eyes open, fully furred, and able to move immediately—and requiring minimal parental care; as in snowshoe hares, deer, porcupines, and many bovids. Contrast altricial.

prehensile. Describing appendages adapted for grasping and seizing by curling or wrapping. Many New World monkeys and the opossums and kinkajous, for example, have prehensile tails.

premolar. Type of cheekteeth of variable structure and function located in front of molars; present in deciduous ("milk") dentition. Primitive mammals have four in each tooth row, above and below. Contrast **molar.**

procumbent. Projecting forward more or less horizontally, as teeth in shrews, horses, and some primates.

promiscuity. Mating system in which animals mate indiscriminately and perhaps often.

pterygoideus. One of three main masticatory muscles of mammals that functions to close the mouth by raising the mandible; important in stabilizing and controlling the movement of the jaw.

reingestion. See **coprophagy.**

rete mirabile. A complex mass of intertwined capillaries specialized for exchange of heat or dissolved substances between countercurrent-flowing blood. Also called *miraculous net.*

reticulum. The second of the four compartments of the stomach of ruminants; a blind-ended sac with honeycomb partitions in its walls.

rostrum (rostral). The preorbital or snout part of the skull of a vertebrate.

r-selection. Selection favoring rapid rates of reproduction and growth, especially among species that specialize in colonizing short-lived, unstable habitats. Contrast **K-selection.**

rumen. The first and largest compartment of the four-part stomach of ruminants.

ruminate (rumination). To chew the cud. See also **foregut fermentation.**

salivary amylase. Potent digestive enzyme produced by the salivary glands.

saltatorial. Adapted for jumping; as in the jumping mouse or kangaroos.

sanguinivorous (sanguinivory). Feeding on a diet of blood; characteristic of vampire bats.

scansorial. Adapted for climbing by means of sharp, curved claws; as in tree squirrels and porcupines.

sciuromorph. Rodents in which the infraorbital foramen is relatively small; squirrel-shaped.

sclerophyll. Type of vegetation characterized by hard, leathery, evergreen foliage that is specially adapted to prevent moisture loss.

semelparity (semelparous). Production of offspring once during a lifetime. Contrast **iteroparity (iteroparous).**

set point. A "reference" temperature in the hypothalamus, analogous to a thermostatic control.

simplex uterus. Type of uterus in which all separation between the uterine horns is lacking; the single uterus opens into the vagina through one cervix. Found in some bats, higher primates, and xenarthrans. Contrast **duplex uterus**.

specialist. Having a lifestyle that involves highly specialized strategies; for example, feeding with one technique on a particular food. Contrast **generalist**.

stapes. Last of the three middle-ear bones (ossicles) found in mammals. In other vertebrates, this is the only ossicle (the columella) in the middle ear. Also called *stirrup*.

submaxillary gland. A small gland under the lower jaw that supplies saliva to the mouth.

subnivean. Beneath a ground cover of snow.

subterranean. Living or occurring beneath the surface of the ground.

temporalis. One of three main masticatory muscles of mammals that functions to close the mouth by raising the mandible. Pronounced in carnivorous mammals, the temporalis assists in holding the jaws closed and aids in the vertical chewing action.

terrestrial. (1) Living on land, as opposed to aquatic. (2) Living primarily on the ground, as opposed to in trees or belowground; for example a coyote.

thermal conductance. Heat loss from the skin to the outside environment.

thermal windows. Bare or sparsely furred areas of certain mammals living in regions characterized by intense solar radiation and high air temperatures, such as guanacos and many desert antelopes. Windows function as sites through which some of the heat gained from solar radiation can be lost by convection and conduction.

thermogenin. A mitochondrial protein responsible for heat production by brown adipose tissue by uncoupling oxidative phosphorylation (separating oxidative processes in the mitochondria from ATP production).

thermoneutral zone. A range in environmental temperatures within which the metabolic rate of an animal is minimal. No energy is required to heat or cool the animal.

thermoregulation The regulation of body temperature in warm-blooded vertebrates (mammals and birds).

torpor. A form of heterothermy in which the body temperature and rate of metabolism are reduced; an adaptation for reducing energy expenditure in periods of cold or food shortage.

total metabolic rate. The total quantity of energy necessary to meet the energy demands of an animal.

tragus. In microchiropteran bats, a projection from the lower margin of the pinnae (ears) that functions in echolocation.

transpirational water loss. See **insensible water loss**.

unicuspid. In shrews, one of the three to five small teeth between the anterior large bicuspid upper incisors and the large cheekteeth.

upper critical temperature. The temperature at which an animal must dissipate heat to maintain a stable internal temperature.

uropatagium. In bats, the membrane between the hind legs that encloses the tail.

vibrissae. Long, stiff, facial whiskers or hairs around the nose or mouth of certain mammals.

volant. Flying; capable of flight.

winter lethargy. Period of winter dormancy in which body temperature decreases only about 5°C to 6°C from euthermy; as in black bears (*Ursus americanus*).

xeric. Of a soil or habitat, dry; having minimal moisture. Contrast **mesic**.

zalambdodont. Having tooth cusps that form a V-shape.

zygomatic arch. Body structure that surrounds and protects the eye and serves as a place of attachment for jaw muscles.

References

Aimi, M., and H. Inagaki. 1988. Grooved lower incisors in flying lemurs. *J. Mammal.* 69:138–140.

Alexander, R. M. 1993. The relative merits of foregut and hindgut fermentation. *J. Zool. Lond.* 231:391–401.

Allen, E. G. 1938. The habits and life history of the eastern chipmunk (*Tamias striatus lysteri*). *Bull. N. Y. State Mus.* 314:1–122.

Allen, J. A. 1877. The influence of physical conditions in the genesis of species. *Radic. Rev.* 1:108–140.

———. 1917. The skeletal characteristics of *Scutiosorex* Thomas. *Bull. Am. Mus. Nat. Hist.* 37:769–784.

Altenbach, J. S. 1989. Prey capture by the fishing bats *Noctilio leporinus* and *Myotis vivesi. J. Mammal.* 70:421–424.

Altringham, J. D. 1996. *Bats: Biology and behaviour.* Oxford: Oxford Univ. Press.

Altringham, J. D., and M. B. Fenton. 2003. Sensory ecology and communication in Chiroptera. In *Bat ecology,* ed. T. H. Kunz and M. B. Fenton, 90–127. Chicago: Univ. of Chicago Press.

Altuna, C. A., L. D. Bacigalupe, and S. Corte. 1998. Food-handling and feces reingestion in *Ctenomys pearsoni* (Rodentia, Ctenomyidae). *Acta Theriol.* 43: 433–437.

Andrews, R. V., and R. W. Belknap. 1986. Bioenergetic benefits of huddling by deer mice (*Peromyscus maniculatus*). *Comp. Biochem. Physiol.* 85A:775–778.

Angerbjörn, A., and J. E. C. Flux. 1995. *Lepus timidus. Mamm. Species* 495:1–11.

Armitage, K. B. 1999. Evolution of sociality in marmots. *J. Mammal.* 80:1–10.

Armstrong, D. M. 2008. *Rocky Mountain mammals: A handbook of mammals of Rocky Mountain National Park and vicinity.* 3d ed. Boulder: Univ. Press of Colorado.

Arnold, W. 1988. Social thermoregulation during hibernation. *J. Comp. Physiol. B Biochem. Syst. Environ. Physiol.* 158:151–156.

Arroyo-Cabrales, J., R. R. Hollander, and J. Knox Jones Jr. 1987. *Choeronycteris mexicana. Mamm. Species* 291:1–5.

Ashton, K. G., M. C. Tracy, and A. de Queiroz. 2000. Is Bergmann's rule valid for mammals? *Am. Nat.* 156:390–415.

Audet, A. M., C. B. Robbins, and S. Larivière. 2002. *Alopex lagopus. Mamm. Species* 713:1–10.

Audubon, J. J., and J. Bachman. 1849. *The quadrupeds of North America.* 2 vols. New York: V. G. Audubon.

Bailey, V. 1924. Breeding, feeding, and other life habits of meadow mice (*Microtus*). *J. Agric. Res.* 27:523–535.

Barnes, B. M. 1989. Freeze avoidance in a mammal: Body temperatures below 0°C in an arctic hibernator. *Science* 244:1593–1595.

———. 1996. Relationships between hibernation and reproduction in male ground squirrels. In *Adaptations to the cold,* ed. F. Geiser, A. Hulbert, and S. Nicol, 71–80. Tenth International Hibernation Symposium. Armidale, NSW, Australia: Univ. of New England Press.

Barnes, B. M., C. Omtzigt, and S. Daan. 1993. Hibernators periodically arouse in order to sleep. In *Life in the cold: Ecological, physiological, and molecular mechanisms,* ed. C. Carey, G. L. Florant, B. A. Wunder and B. Horwitz, 555–558. Boulder, CO: Westview Press.

Barry, R. E. 2009. Between a rock and a hyrax. *Nat. Hist.* 118:30–35

Barry, R. E., and J. Shoshani. 2000. *Heterohyrax brucei. Mamm. Species* 645:1–7.

Bartholomew, G. A. 1982. Body temperature and energy metabolism. In *Animal physiology: Principles and adaptations,* 4th ed., ed. M. S. Gordon, G. A. Bartholomew, A. D. Grinnell, C. B. Jørgensen, and F. N. White, 333–406. New York: Macmillan.

Bat Conservation International. 2006. A new world record. *Bats,* winter, 14.

Baudinette, R. V., S. K. Churchhill, K. A. Christian, J. E. Nelson, and P. J. Hudson. 2000. Energy, water balance and the roost microenvironment in three Australian cave-dwelling bats (Microchiroptera). *J. Comp. Physiol. B* 170:439–446.

Bazin, R. C., and R. A. MacArthur. 1992. Thermal benefits of huddling in the muskrat (*Ondatra zibethicus*). *J. Mammal.* 73:559–564.

Bearder, S. K., and R. D. Martin. 1980. Acacia gum and its use by bush-babies, *Galago senegalensis* (Primates: Lorisidae). *Int. J. Primatol.* 1:103–128.

Beidleman, R. G., and W. A. Weber. 1958. Analysis of a pika hay pile. *J. Mammal.* 39:599–600.

Bennett, A. F., R. B. Huey, H. J. Alder, and K. A. Nagy. 1984. The parasol tail and thermoregulatory behavior of the cape ground squirrel, *Xerus inauris. Physiol. Zool.* 57:57–62.

Benstead, J. P., K. H. Barnes, and C. M. Pringle. 2001. Diet, activity patterns, foraging movement, and responses to deforestation of the aquatic tenrec *Limno-*

gale mergulus (Lipotyphla: Tenrecidae) in eastern Madagascar. *J. Zool.* 254: 119–129.

Betke, M., D. H. Hirsh, N. C. Makris, G. F. McCracken, M. Procopio, N. I. Hristov, S. Tang, A. Bagchi, J. D. Reichard, J. W. Horn, S. Crampton, C. J. Cleveland, and T. H. Kunz. 2008. Thermal imaging reveals significantly smaller Brazilian free-tailed bat colonies than previously estimated. *J. Mammal.* 89: 18–24.

Blehert, D. S., A. C. Hicks, M. Behr, C. U. Meteyer, B. M. Berlowski-Zier, E. L. Buckles, J. T. H. Coleman, S. R. Darling, A. Gargas, R. Niver, J. C. Okoniewski, R. J. Rudd, and W. B. Stone. 2008. Bat white-nose syndrome: An emerging fungal pathogen? *Science* 30 October 2008. 10.1126/science.1163874.

Blood, B. R., and M. K. Clark. 1998. *Myotis vivesi. Mamm. Species* 588:1–5.

Bodemer, C. W. 1968. *Modern embryology.* New York: Holt, Rinehart and Winston.

Bogdanowicz, W. 1994. *Myotis daubentonii. Mamm. Species* 475:1–9.

Bogdanowicz, W., M. B. Fenton, and K. Daleszczyk. 1999. The relationship between echological calls, morphology, and diet in insectivorous bats. *J. Zool. Lond.* 247:381–393.

Bordignon, M. O. 2006. Diet of the fishing bat *Noctilio leporinus* (Linnaeus) (Mammalia, Chiroptera) in a mangrove area of southern Brazil. *Rev. Brasil. Zool.* 23:256–260.

Bourliere, F. 1975. Mammals, small and large: The ecological implications of size. In *Small mammals: Their productivity and population dynamics,* ed. F. B. Golley, K. Petrusewicz, and L. Ryszkowski, 1–8. IBP no. 5. Cambridge: Cambridge Univ. Press.

Boyer, B. B., and B. M. Barnes. 1999. Molecular and metabolic aspects of mammalian hibernation. *Bioscience* 49:713–724.

Brack, V. Jr., and J. O. Whitaker Jr. 2001. Foods of the northern myotis, *Myotis septentrionalis,* from Missouri and Indiana, with notes on foraging. *Acta Chiropterol.* 3:203–210.

Bradbury, J. W. 1977. Lek mating behavior in the hammer-headed bat. *Z. Tierpsychol.* 45:225–255.

Bradshaw, F. J., and S. D. Bradshaw. 2001. Maintenance nitrogen requirement of an obligate nectarivore, the honey possum, *Tarsipes rostratus. J. Comp. Physiol. B* 171:59–67.

Bradshaw, G. V. R. 1962. Reproductive cycle of the California leaf-nosed bat, *Macrotus californicus. Science* 136:645–646.

Bradshaw, S. D., K. D. Morris, and F. J. Bradshaw. 2001. Water and electrolyte homeostasis and kidney function of desert-dwelling marsupial wallabies in western Australia. *J. Comp. Physiol. B* 171:23–32.

Braude, S. 2000. Dispersal and new colony formation in wild naked mole-rats: Evidence against inbreeding as the system of mating. *Behav. Ecol.* 11:7–12.

Brett, R. A. 1986. The ecology and behavior of the naked mole-rat (*Heterocephalus glaber* Rüppell) (Rodentia: Bathyergidae). Ph.D. diss., Univ. of London.

Brice, P. H., G. C. Grigg, L. A. Beard, and J. A. Donovan. 2002. Patterns of activity and inactivity in echidnas (*Tachyglossus aculeatus*) free-ranging in a hot dry climate: Correlates with ambient temperature, time of day and season. *Aust. J. Zool.* 50:461–475.

Bronner, G. N., and P. D. Jenkins. 2005. Order Afrosoricida. In *Mammal species of the world: A taxonomic and geographic reference,* 3d ed., vol. 2, ed. D. E. Wilson and D. M. Reeder, 71–81. Baltimore: Johns Hopkins Univ. Press.

Brooke, A. P. 1994. Diet of the fishing bat, *Noctilio leporinus* (Chiroptera: Noctilionidae). *J. Mammal.* 75:212–218.

Brosset, A. 1989. Camouflage chez le yapock *Chironectes minimus. Rev. Ecol.* 44: 279–281.

Brown, J. H., and R. C. Lasiewski. 1972. Metabolism of weasels: The cost of being long and thin. *Ecology* 53:939–943.

Brown, J. H., O. J. Reichman, and D. W. Davidson. 1979. Granivory in desert ecosystems. *Annu. Rev. Ecol. Syst.* 10:201–227.

Brown, K. J., and C. T. Downs. 2005. Seasonal behavioural patterns of free-living rock hyrax (*Procavia capensis*). *J. Zool.* 265:311–326.

———. 2006. Seasonal patterns in body temperature of free-living rock hyrax (*Procavia capensis*). *Comp. Biochem. Physiol. A Mol. Int. Physiol.* 143:42–49.

Bruns, V., H. Burda, and M. J. Ryan. 1989. Ear morphology of the frog-eating bat (*Trachops cirrhosus,* Family: Phyllostomidae): Apparent specializations for low-frequency hearing. *J. Morphol.* 199:103–118.

Buech, R. R., D. J. Rugg, and N. L. Miller. 1989. Temperature in beaver lodges and bank dens in a near-boreal environment. *Can. J. Zool.* 67:1061–1066.

Buffenstein, R., and S. Yahav. 1991. Is the naked mole-rat, *Heterocephalus glaber,* an endothermic yet poikilothermic mammal? *J. Therm. Biol.* 16:227–232.

Burnett, C. D. 1983. Geographic and climate correlates on morphological variation in *Eptesicus fuscus. J. Mammal.* 64:437–444.

Calder, W. A. 1969. Temperature relations and under water endurance of the smallest homeothermic diver, the water shrew. *Comp. Biochem. Physiol.* 30A:1075–1082.

Carleton, M. D., and G. G. Musser. 2005. Order Rodentia. In *Mammal species of the world: A taxonomic and geographic reference,* 3d ed., vol. 2, ed. D. E. Wilson and D. M. Reeder, 745–752. Baltimore: Johns Hopkins Univ. Press.

Catania, K. C. 1995. A comparison of the Eimer's organs of three North American moles: The hairy-tailed mole (*Parascalops breweri*), the star-nosed mole (*Condylura cristata*), and the eastern mole (*Scalopus aquaticus*). *J. Comp. Neurol.* 354:150–160.

———. 1999. A nose that looks like a hand and acts like an eye: The unusual mechanosensory system of the star-nosed mole. *J. Comp. Physiol.* A 185:367–372.

———. 2000. Epidermal sensory organs of moles, shrew-moles, and desmans: A study of the family Talpidae with comment on the function and evolution of Eimer's organ. *Brain Behav. Evol.* 56:146–174.

———. 2006. Underwater "sniffing" by semi-aquatic mammals. *Nature* 444:1024–1025.

Catania, K. C., J. F. Hare, and K. L. Campbell. 2008. Water shrews detect movement, shape, and smell to find prey underwater. *Proc. Natl. Acad. Sci. USA* 105:571–576.

Catania, K. C., and J. H. Kaas. 1996. The unusual nose and brain of the star-nosed mole. *Bioscience* 46:578–586.

Chan, L. K. 1995. Extrinsic lingual musculature of two pangolins (Pholidota: Manidae). *J. Mammal.* 76:472–480.

Chappell, M. A., and G. A. Bartholomew. 1981. Standard operative temperatures and thermal energetics of the antelope ground squirrel *Ammospermophilus leucurus*. *Physiol. Zool.* 54:81–93.

Charles-Dominique, P., and J.-J. Petter. 1980. Ecology and social life of *Phaner furcifer*. In *Nocturnal Malagasy primates*, ed. P. Charles-Dominique, H. M. Cooper, A. Hladik, C. M. Hladik, E. Pages, G. F. Pariente, A. Petter-Rousseaux, A. Schilling, and J. J. Petter, 75–95. New York: Academic Press.

Chitty, D. 1996. *Do lemmings commit suicide? Beautiful hypotheses and ugly facts.* New York: Oxford Univ. Press.

Christiansen, P., and S. Wroe. 2007. Bite forces and evolutionary adaptations to feeding ecology in carnivores. *Ecology* 88:347–358.

Churchfield, S. 1982. The influence of temperature on the activity and food consumption of the common shrew. *Acta Theriol.* 27:295–304.

———. 1990. *The natural history of shrews.* Ithaca, NY: Cornell Univ. Press.

Claridge, A. W., and S. J. Cork. 1994. Nutritional value of hypogeal fungal sporocarps for the long-nosed potoroo (*Potorous tridactylus*), a forest-dwelling mycophagous marsupial. *Aust. J. Zool.* 42:701–710.

Clutton-Brock, J., ed. 1989. *The walking larder: Patterns of domestication, pastoralism, and predation.* London: Unwin Hyman.

Clutton-Brock, T. H., ed. 1977. *Primate ecology: Studies of feeding and ranging behaviour in lemurs, monkeys, and apes.* London: Academic Press.

Clutton-Brock, T. H., P. N. M. Brotherton, M. J. O'Riain, A. S. Griffin, D. Gaynor, R. Kansky, L. Sharpe, and G. M. McIlrath. 2001. Contributions to cooperative rearing in meerkats. *Anim. Behav.* 61:705–710.

Cockburn, A. 1997. Living slow and dying young: Senescence in marsupials. In *Marsupial biology: Recent research, new perspectives*, ed. N. Saunders and L. Hinds, 163–174. Sydney: Univ. of New South Wales Press.

Coles, R. W. 1969. Thermoregulatory function of the beaver tail. *Am. Zool.* 9:203a.

Conner, D. A. 1983. Seasonal changes in activity patterns and the adaptive value of haying in pikas (*Ochotona princeps*). *Can. J. Zool.* 61:411–416.

Constantine, D. C. 1967. Activity patterns of the Mexican free-tailed bat. *Univ. N. M. Publ. Biol.* 7:1–79.

Cook, J. A., S. Anderson, and T. L. Yates. 1990. Note on Bolivian mammals. 6: The genus *Ctenomys* (Rodentia, Ctenomyidae) in the highlands. *Am. Mus. Novitates* 2980:1–27.

Corbett, L. K. 1995. *The dingo in Australia and Asia.* Sydney: New South Wales Univ. Press.

Cork, S. J., and G. J. Kenagy. 1989. Rates of gut passage and retention of hypogeous fungal spores in two forest-dwelling rodents. *J. Mammal.* 70:512–519.

Coulson, T., and A. Malo. 2008. Case of the absent lemmings. *Nature* 456:43–44.

Cramer, M. J., M. R. Willig, and C. Jones. 2001. *Trachops cirrhosus. Mamm. Species* 656:1–6.

Cranford, J. A. 1978. Hibernation in the western jumping mouse (*Zapus princeps*). *J. Mammal.* 59:496–509.

Crawford, J. C., Z. Liu, T. A. Nelson, C. K. Nielsen, and C. K. Bloomquist. 2008. Microsatellite analysis of mating and kinship in beavers (*Castor canadensis*). *J. Mammal.* 89:575–581

Cullinane, D. M., and D. Aleper. 1998. The functional and biomechanical modifications of the spine of *Scutisorex somereni,* the hero shrew: Spinal musculature. *J. Zool. Lond.* 244:453–458.

Cullinane, D. M., D. Aleper, and J. E. A. Bertram. 1998. The functional and biomechanical modifications of the spine of *Scutiosorex somereni,* the hero shrew: Skeletal scaling relationships. *J. Zool. Lond.* 244:447–452.

Cullinane, D. M., and J. E. A. Bertram. 2000. The mechanical behaviour of a novel mammalian intervertebral joint. *J. Anat.* 197:627–634.

Cutright, W. J., and T. McKean. 1979. Countercurrent blood vessel arrangement in beaver (*Castor canadensis*). *J. Morphol.* 161:169–176.

Cypher, B. L. 2003. Foxes. In *Wild mammals of North America: Biology, management, and conservation,* ed. G. A. Feldhamer, B. C. Thompson, and J. A. Chapman, 511–546. Baltimore: Johns Hopkins Univ. Press.

Daniel, J. C. Jr. 1970. Dormant embryos of mammals. *Bioscience* 20:411–415.

Dausmann, K. H., J. Glos, J. U. Ganahorn, and G. Heldmaier. 2004. Hibernation in a tropical primate. *Nature* 429:825–826.

Dawson, T. J. 1995. *Kangaroos: Biology of the largest marsupials.* Ithaca, NY: Comstock.

Dearing, M. D. 1997. The function of hay piles of pikas (*Ochotona princeps*). *J. Mammal.* 78:1156–1163.

Degen, A. A. 1997. *Ecophysiology of small desert mammals.* Berlin: Springer-Verlag.

Delany, M. J. 1974. *The ecology of small mammals*. Institute of Biology's Studies in Biology, no. 51. New York: Crane, Russak.

De Oliveira, T. G. 1998. *Leopardus wiedii. Mamm. Species* 579:1–6.

D'Havé, H., J. Scheirs, R. Vergagen, and W. DeCoen. 2005. Gender, age and seasonal dependent self-anointing in the European hedgehog *Erinaceus europaeus. Acta Theriol.* 50:167–173.

Dietz, M., and E. K. V. Kalko. 2006. Seasonal changes in daily torpor patterns of free-ranging female and male Daubenton's bats (*Myotis daubentonii*). *J. Comp. Physiol. B* 176:223–231.

Dijkgraaf, S. 1952. Spallanzani und die Fledermäuse. *Experientia* 8:205–216.

Dokuchaev, N. E. 1989. Population ecology of *Sorex* shrews in north-east Siberia. *Ann. Zool. Fennici* 26:371–379.

Dondini, G., and S. Vergari. 2000. Carnivory in the greater noctule bat (*Nyctalus lasiopterus*) in Italy. *J. Zool. Lond.* 251:233–236.

Dwyer, P. D. 1970. Foraging behaviour of the Australian large-footed *Myotis* (Chiroptera). *Mammalia* 34:76–80.

Dyck, A. P., and R. A. MacArthur. 1993. Seasonal variation in the microclimate and gas composition of beaver lodges in a boreal environment. *J. Mammal.* 74:180–188.

Elbroch, M. 2006. *Animal skulls: A guide to North American species*. Mechanicsburg, PA: Stackpole Books.

Elton, C. 1924. Periodic fluctuations in the numbers of animals: Their causes and effects. *Br. J. Exp. Biol.* 2:119–163.

Elton, C. S. 1927. *Animal ecology*. London: Sidgwick and Jackson.

Emmons, L. H. 1991. Frugivory in treeshrews (*Tupaia*). *Am. Nat.* 138:642–649.

———. 1997. *Neotropical rainforest mammals: A field guide*. 2d ed. Chicago: Univ. of Chicago Press.

———. 2000. *Tupai: A field study of Bornean treeshrews*. Berkeley: Univ. of California Press.

Faulkes, C. G., and D. H. Abbott. 1993. Evidence that primer pheromones do not cause social suppression of reproduction in male and female naked mole-rats (*Heterocephalus glaber*). *J. Reprod. Fertil.* 99:225–230.

Feldhamer, G. A., L. C. Drickamer, S. H. Vessey, J. F. Merritt, and C. Krajewski. 2007. *Mammalogy: Adaptation, diversity, ecology*. 3d ed. Baltimore: Johns Hopkins Univ. Press.

Finerty, J. P. 1980. *The population ecology of cycles in small mammals: Mathematical theory and biological fact*. New Haven, CT: Yale Univ. Press.

Fisher, D. O., and C. R. Dickman. 1993. Body-size prey-size relationships in insectivorous marsupials—tests of three hypotheses. *Ecology* 74:1871–1883.

Flannery, T. F., and C. P. Groves. 1998. A revision of the genus *Zaglossus* (Monotremata, Tachyglossidae), with description of new species and subspecies. *Mammalia* 62:367–396.

Fleming, T. H. 1971. *Artibeus jamaicensis*: Delayed embryonic development in a neotropical bat. *Science* 171:402–404.

———. 1979. Life history strategies. In *Ecology of small mammals,* ed. M. Stoddart, 1–61. London: Chapman and Hall.

———. 1982. Foraging strategies of plant-visiting bats. In *Ecology of bats,* ed. T. H. Kunz, 287–325. New York: Plenum.

Flux, J. E. 1970. Colour change of mountain hares (*Lepus timidus scoticus*) in northeast Scotland. *J. Zool.* 162:345–358.

Fogel, R., and J. M. Trappe. 1978. Fungus consumption (mycophagy) by small animals. *Northwest Sci.* 52:1–31.

Fooden, J. 1997. Tail length variation in *Macaca fascicularis* and *M. mulatta*. *Primates* 38:221–231.

Fooden, J., and G. H. Albrecht. 1999. Tail-length evolution In *fascicularis*-group macaques (Cercopithecidae: *Macaca*). *Int. J. Primatol.* 20:431–440.

Foster, W. K., W. Caton, J. Thomas, S. Cox, and D. A. Taggart. 2008. Timing of births and reproductive success in captive red-tailed phascogales, *Phascogale calura*. *J. Mammal.* 89:1136–1144.

Frafjord, K. 1992. Denning behaviour and activity of arctic fox *Alopex lagopus* pups: Implications of food availability. *Popular Biol.* 12:707–712.

Freckleton, R. P., P. H. Harvey, and M. Pagel. 2003. Bergmann's rule and body size in mammals. *Am. Nat.* 161:821–825.

Freeman, P. W. 1988. Frugivorous and animalivorous bats (Microchiroptera): Dental and cranial adaptations. *Biol. J. Linnean Soc.* 33:249–272.

French, A. R. 1985. Allometries of the durations of torpid and euthermic intervals during mammalian hibernation: A test of the theory of metabolic control of the timing of changes in body temperature. *J. Comp. Physiol. B* 156:13–19.

———. 1993. Physiological ecology of the Heteromyidae: Economics of energy and water utilization. In *Biology of the Heteromyidae,* ed. H. H. Genoways and J. H. Brown, 509–538. Spec. pub. Am. Soc. Mammal., vol. 10. Lawrence, KS: American Society of Mammalogists.

Freudenberger, D. O., I. R. Wallis, and I. D. Hume. 1989. Digestive adaptations of kangaroos, wallabies, and rat-kangaroos. In *Kangaroos, wallabies, and rat-kangaroos,* ed. G. Grigg, P. Jarman, and I. D. Hume, 179–189. Chipping Norton, NSW, Australia: Surrey Beatty.

Frost, H. C., W. B. Krohn, and C. R. Wallace. 1997. Age-specific reproductive characteristics in fishers. *J. Mammal.* 78:598–612.

Gaisler, J. 1979. Ecology of bats. In *Ecology of small mammals,* ed. D. M. Stoddart, 281–342. London: Chapman and Hall.

Galambos, R. 1942. The avoidance of obstacles by flying bats: Spallanzani's ideas (1794) and later theories. *Isis* 34:132–140.

Galliez, M., M. D. S. Leite, T. L. Queiroz, and F. A. S. Fernandez. 2009. Ecology of the water opossum *Chironectes minimus* in Atlantic forest streams of southeastern Brazil. *J. Mammal.* 90:93–103.

Gardner, A. L. 1977. Feeding habits. In *Biology of bats of the New World family Phyllostomatidae,* pt. II, ed. R. J. Baker, J. Knox Jones Jr., and D. C. Carter, 293–350. Museum of Texas Tech Univ. spec. pub. no. 13. Lubbock: Museum of Texas Tech Univ.

Geffen, E. 1994. *Vulpes cana. Mamm. Species* 462:1–4.

Geffen, E., R. Hefner, D. W. Macdonald, and M. Ucko. 1992. Diet and foraging behavior of Blandford's foxes, *Vulpes cana,* in Israel. *J. Mammal.* 73:395–402.

Geiser, F. 2004. Metabolic rate and body temperature reduction during hibernation and torpor. *Annu. Rev. Physiol.* 66:239–274.

Geiser, F., and R. V. Baudinette. 1987. Seasonality of torpor and thermoregulation in three dasyurid marsupials. *J. Comp. Physiol.* 157B:335–344.

Geiser, F., and T. Ruff. 1995. Hibernation versus daily torpor in mammals and birds: Physiological variable and classification of torpor patterns. *Physiol. Zool.* 68: 935–966.

Geiser, F., H. S. Sink, B. Stahl, I. M. Mansergh, and L. S. Broome. 1990. Differences in the physiological response to cold in the wild and laboratory-bred mountain pygmy possums *Burramys parvus* (Marsupialia). *Aust. Wildl. Res.* 17:535–539.

Geist, V. 1987. Bergmann's rule is invalid. *Can. J. Zool.* 65:1035–1038.

Genoud, M. 1988. Energetic strategies of shrews: Ecological constraints and evolutionary implications. *Mammal Rev.* 18:173–193.

Gilg, O. 2002. The summer decline of the collared lemming, *Dicrostonyx groenlandicus,* in high arctic Greenland. *Oikos* 99:499–510.

Gilg, O., B. Sittler, B. Sabard, A. Hurstel, R. Sane, P. Delattre, and L. Hanski. 2006. Functional and numerical responses of four lemming predators in high arctic Greenland. *Oikos* 113:193–216.

Gloger, C. L. 1833. *Das Abändern der Vögel durch Einfluss des Klimas.* Breslau, Germany.

Goldingay, R. L. 1990. The foraging behaviour of a nectar feeding marsupial *Petaurus australis. Oecologia* 85:191–199.

Goldingay, R. L., and R. P. Kavangah. 1991. The yellow-bellied glider: A review of its ecology and management considerations. In *Conservation of Australia's forest fauna,* ed. D. Lunney, 365–375. Sydney, Australia: Royal Zoological Society.

Goldingay, R. L., and J. Scheibe, eds. 2000. *Biology of gliding mammals.* Fürth, Germany: Filander-Verlag.

Golley, F. B., K. Petrusewicz, and L. Ryszkowski, eds. 1975. *Small mammals: Their productivity and population dynamics.* IBP no. 5. Cambridge: Cambridge Univ. Press.

Gordon, M. S., G. A. Bartholomew, A. D. Grinnell, C. B. Jørgensen, and F. N. White, eds. 1982. *Animal physiology: Principles and adaptations.* 4th ed. New York: Macmillan.

Gorman, M. L., and R. D. Stone. 1990. *The natural history of moles*. Ithaca, NY: Comstock.

Grant, T. 1995. *The platypus*. Sydney: Univ. of New South Wales Press.

Green, K., and W. Osborne. 1994. *Wildlife of the Australian snow country: A comprehensive guide to alpine fauna*. Chatswood, NSW: Reed Book Australia.

Greenhall, A. M., G. Joermann, and U. Schmidt. 1983. *Desmodus rotundus. Mamm. Species* 202:1–6.

Greenhall, A. M., and U. Schmidt, eds. 1988. *Natural history of vampire bats*. Boca Raton, FL: CRC Press.

Greenhall, A. M., U. Schmidt, and G. Joermann. 1984. *Diphylla ecaudata. Mamm. Species* 227:1–3.

Greenhall, A. M., and W. A. Schutt Jr. 1996. *Diaemus youngi. Mamm. Species* 533:1–7.

Griffin, D. R. 1958. *Listening in the dark*. New Haven, CT: Yale Univ. Press.

Griffiths, M. 1968. *Echidnas*. Oxford: Pergamon Press.

Griffiths, M., R. T. Wells, and D. J. Barrie. 1991. Observations on the skulls of fossil and extant echidnas (Monotremata: Tachyglossidae). *Aust. Mammal.* 14:87–101.

Grigg, G. C., L. A. Beard, and M. L. Augee. 1989. Hibernation in a monotreme, the echidna *Tachyglossus aculeatus. Comp. Biochem. Physiol.* 92A:609–612.

———. 2004. The evolution of endothermy and its diversity in mammals and birds. *Physiol. Biochem. Zool.* 77:982–997.

Grigg, G. C., L. A. Beard, T. R. Grant, and M. L. Augee. 1992. Body temperature and diurnal activity pattern in the platypus, *Ornithorhynchus anatinus*, during winter. *Aust. J. Zool.* 40:135–142.

Grinnell, J., and R. T. Orr. 1934. Systematic review of the *californicus* group of the rodent genus *Peromyscus. J. Mammal.* 15:210–220.

Grojean, R. E., J. A. Suusa, and M. C. Henry. 1980. Utilization of solar radiation by polar animals: An optical model for pelts. *Appl. Optics* 19:339–346.

Groves, C. P. 2005. Order Diprotodontia. In *Mammal species of the world: A taxonomic and geographic reference*, 3d ed., vol. 1, ed. D. E. Wilson and D. M. Reeder, 43–70. Baltimore: Johns Hopkins Univ. Press.

Gunderson, H. L. 1976. *Mammalogy*. New York: McGraw-Hill.

Gunson, J. R., and R. R. Bjorge. 1979. Winter denning of the striped skunk in Alberta. *Can. Field-Naturalist* 93:252–258.

Gurnell, J. 1987. *The natural history of squirrels*. London: Christopher Helm.

Hainsworth, F. R. 1981. *Animal physiology: Adaptations in function*. Reading, MA: Addison-Wesley.

Hamilton, J. L., and R. M. R. Barclay. 1995. Patterns of daily torpor and day roost selection by male and female big brown bats (*Eptesicus fuscus*). *Can. J. Zool.* 72:744–749.

Hamilton, W. J. Jr. 1937. The biology of microtine cycles. *J. Agric. Res.* 54:779–790.

Hanski, I., L. Hansson, and H. Henttonen. 1991. Specialist predators, generalist predators, and the microtine rodent cycle. *J. Anim. Ecol.* 60:353–367.

Harlow, H. J., T. Lohuis, R. C. Anderson-Sprecher, and T. D. I. Beck. 2004. Body surface temperature of hibernating black bears may be related to periodic muscle activity. *J. Mammal.* 85:414–419.

Havera, S. P., and C. M. Nixon. 1978. Geographic variation of Illinois gray squirrels. *Am. Midl. Nat.* 100:396–407.

Hays, J. P. 2001. Mass-specific and whole-animal metabolism are not the same concept. *Physiol. Biochem. Zool.* 74:147–150.

Hays, J. P., and J. S. Shonkwiler. 1996. Analyzing mass-independent data. *Physiol. Zool.* 69:974–980.

Hayssen, V., A. van Tienhoven, and A. van Tienhoven. 1993. *Asdell's patterns of mammalian reproduction.* Ithaca, NY: Comstock.

Hayward, J. S., and C. P. Lyman. 1967. Nonshivering heat production during arousal from hibernation and evidence for the contribution of brown fat. In *Mammalian hibernation III,* ed. K. C. Fisher, A. R. Dawe, C. P. Lyman, E. Schönbaum, and F. E. South, 346–355. Edinburgh: Oliver and Boyd,.

Heideman, P. D. 1989. Delayed development in Fischer's pygmy fruit bat, *Haplonycteris fischeri,* in the Philippines. *J. Reprod. Fertil.* 85:363–382.

Heideman, P. D., J. A. Cummings, and L. R. Heaney. 1993. Reproductive timing and early embryonic development in an Old World fruit bat, *Otopteropus cartilagonodus* (Megachiroptera). *J. Mammal.* 74:621–630.

Heideman, P. D., and K. S. Powell. 1998. Age-specific reproductive strategies and delayed embryonic development in an Old World fruit bat, *Ptenochirus jagori.* *J. Mammal.* 79:295–311.

Helgen, K. M. 2005. Order Scandentia. In *Mammal species of the world: A taxonomic and geographic reference,* 3d ed., vol. 1, ed. D. E. Wilson and D. M. Reeder, 104–109. Baltimore: Johns Hopkins Univ. Press.

Heller, H. C., D. A. Grahn, L. Trachsel, and J. E. Larkin. 1993. What is a bout of hibernation? In *Life in the cold: Ecological, physiological, and molecular mechanisms,* ed. C. Carey, G. L. Florant, B. A. Wunder, and B. Horwitz, 253–264. Boulder, CO: Westview Press.

Heller, H. C., and N. F. Ruby. 2004. Sleep and circadian rhythms in mammalian torpor. *Annu. Rev. Physiol.* 66:275–289.

Henshaw, R. E., L. S. Underwood, and T. M. Casey. 1972. Peripheral thermoregulation: Foot temperature in two arctic canines. *Science* 175:988–990.

Herbst, M., and N. C. Bennett. 2006. Burrow architecture and burrowing dynamics of the endangered Namaqua dune mole rat (*Bathyergus janetta*) (Rodentia: Bathyergidae). *J. Zool. Lond.* 270:420–428.

Hermanson, J. W., and T. J. O'Shea. 1983. *Antrozous pallidus. Mamm. Species* 213:1–8.

Hess, J. 2004. A population genetic study of the eusocial naked mole-rat (*Heterocephalius glaber*). Ph.D. diss., Washington Univ., St. Louis, MO.

Heusner. A. A. 1991. Size and power in mammals. *J. Exp. Biol.* 160:25–54.

Hill, J. E., and J. D. Smith. 1992. *Bats: A natural history.* Austin: Univ. of Texas Press.

Hill, R. W., D. P. Christian, and J. H. Veghte. 1980. Pinna temperature in exercising jackrabbits. *J. Mammal.* 61:30–38.

Hill, R. W., and G. A. Wyse. 1989. *Animal physiology.* 2d ed. New York: Harper and Row.

Himms-Hagen, J. 1985. Brown adipose tissue metabolism and thermogenesis. *Annu. Rev. Nutr.* 5:69–94.

Hirakawa, H. 2001. Coprophagy in leporids and other mammalian herbivores. *Mammal Rev.* 31:61–80.

Hladik, C. M., P. Charles-Dominique, and J.-J. Petter. 1980. Feeding strategies of five nocturnal prosimians in the dry forest of the west coast of Madagascar. In *Nocturnal Malagasy primates,* ed. P. Charles-Dominique, H. M. Cooper, A. Hladik, C. M. Hladik, E. Pages, G. F. Pariente, A. Petter-Rousseaux, and A. Schilling., 41–73. New York: Academic Press.

Hoeck, H. N. 1982. Population dynamics, dispersal and genetic isolation in two species of hyrax (*Heterohyrax brucei* and *Procavia johnstoni*) on habitat islands in the Serengeti. *Z. Tierpsychol.* 59:177–210.

Holland, R. A., D. A. Waters, and J. M. V. Rayner. 2004. Echolocation signal structure in the megachiropteran bat *Rousettus aegyptiacus* Geoffroy 1810. *J. Exp. Biol.* 207:4361–4369.

Honeycutt, R. L. 1992. Naked mole-rats. *Am. Sci.* 80:43–53.

Hood, C. S., and J. Knox Jones Jr. 1984. *Noctilio leporinus. Mamm. Species* 216:1–7.

Howell, A. H. 1920. The Florida spotted skunk as an acrobat. *J. Mammal.* 1:88.

Hughes, R. L. 1984. Structural adaptations of the eggs and the fetal membranes of monotremes and marsupials for respiration and metabolic exchange. In *Respiration and metabolism of embryonic vertebrates,* ed. R. S. Seymour, 389–421. Dordrecht, Netherlands: Junk.

Hume, I. D. 1989. Optimal digestive strategies in mammalian herbivores. *Physiol. Zool.* 62:1145–1163.

Hunter, L. 2005. *Cats of Africa: Behavior, ecology, and conservation.* Baltimore: Johns Hopkins Univ. Press.

Hurst, R. N., and J. E. Wiebers. 1967. Minimum body temperature extremes in the little brown bat, *Myotis lucifugus. J. Mammal* 48:465.

Hutterer, R. 2005. Order Soricomorpha. In *Mammal species of the world: A taxonomic and geographic reference,* 3d ed., vol. 1, ed. D. E. Wilson and D. M. Reeder, 220–311. Baltimore: Johns Hopkins Univ. Press.

Hwang, Y. T., S. Larivière, and F. Messier. 2007. Energetic consequences and ecological significance of heterothermy and social thermoregulation in striped skunks (*Mephitis mephitis*). *Physiol. Biochem. Zool.* 80:138–145.

Hyvärinen, H. 1994. Brown fat and the winter of shrews. In *Advances in the biology of shrews,* ed. J. F. Merritt, G. L. Kirkland Jr., and R. K. Rose, 259–266. Carnegie Museum of Natural History spec. pub. no. 18. Pittsburgh: Carnegie Museum of Natural History.

Ibanez, C., J. Juste, J. L. Garcia-Mudarra, and P. T. Agirre-Mendi. 2001. Bat predation on nocturnally migrating birds. *Proc. Natl. Acad. Sci. USA* 98:9700–9702.

Ims, R. A., J.-A. Henden, and S. T. Killengreen. 2008. Collapsing population cycles. *Trends Ecol. Evol.* 23:79–86.

Ingles, L. G. 1961. Reingestion in the mountain beaver. *J. Mammal.* 42:411–412.

Ivanter, E. V. 1994. The structure and adaptive peculiarities of pelage in soricine shrews. In *Advances in the biology of shrews,* ed. J. F. Merritt, G. L. Kirkland Jr., and R. K. Rose, 441–454. Carnegie Museum of Natural History spec. pub. no. 18. Pittsburgh: Carnegie Museum of Natural History.

Jackson, H. H. T. 1961. *Mammals of Wisconsin.* Madison: Univ. of Wisconsin Press.

James, F. C. 1970. Geographic size variation in birds and its relationship to climate. *Ecology* 51:365–390.

Jarvis, J. U. M. 1981. Eusociality in a mammal: Cooperative breeding in naked mole-rat colonies. *Science* 212:571–573.

Jarvis, J. U. M., and P. W. Sherman. 2002. *Heterocephalis glaber. Mamm. Species* 706:1–9.

Jett, D. A., and J. D. Nichols. 1987. Density fluctuations in a meadow vole population at the Patuxent Wildlife Research Center. *Md. Nat.* 31:41–43.

Johnson, E. 1984. Seasonal adaptive coat changes in mammals. *Acta Zool. Fennica* 171:7–12.

Johnson, E., and J. Hornby. 1980. Age and seasonal coat changes in long haired and normal fallow deer (*Dama dama*). *J. Zool.* 192:501–509.

Jones, G., and J. M. V. Rayner. 1988. Flight performance, foraging tactics and echolocation in free-living Daubenton's bats *Myotis daubentonii* (Chiroptera: Vespertilionidae). *J. Zool. Lond.* 215:113–132.

———. 1991. Flight performance, foraging tactics and echolocation in the trawling insectivorous bat *Myotis adversus* (Chiroptera: Vespertilionidae). *J. Zool. Lond.* 225:393–412.

Joshi, A. R., D. L. Garshelis, and J. L. D. Smith. 1997. Seasonal and habitat-related diets of sloth bears in Nepal. *J. Mammal.* 78:584–597.

Jurine, L. 1798. Extrait des expériences de Jurine sur les chauves-souris qu'on a privé(es) de la vue. *J. Phys.* 46:145–148 (English trans. in *Philos. Mag.* 1:136–140).

Justo, E. R., L. J. M. De Santis, and M. S. Kin. 2003. *Ctenomys talarum. Mamm. Species* 730:1–5.

Kalko, E. K. V., D. Friemel, C. O. Handley Jr., and H.-U. Schnitzler. 1999. Roosting and foraging behavior of two neotropical gleaning bats, *Tonatia silvicola* and *Trachops cirrhosus* (Phyllostomidae). *Biotropica* 31:344–353.

Kalko, E. K. V., and H.-U. Schnitzler. 1998. How echolocating bats approach and acquire food. In *Bat biology and conservation,* ed. T. H. Kunz and P. A. Racey, 197–204. Washington, DC: Smithsonian Institution Press.

Kallen, F. C. 1977. The cardiovascular systems of bats: Structure and function. In *Biology of bats,* ed. W. A. Wimsatt, 290–483. New York: Academic Press.

Karaseva, E. V., M. Telitsyn, V. A. Lapshov, and Y. V. Okhotsky. 1971. Study of the land vertebrates of Central Yamal. *Byulleten Moskovskogo obshchestva ispytatelei prirody* 76:22–32.

Kausrud, K. L., A. Mysterud, H. Steen, J. O. Vik, E. Østbye, B. Cazelles, E. Framstad, A. M. Eikeset, I. Mysterud, T. Solhøy, and N. C. Stenseth. 2008. Linking climate change to lemming cycles. *Nature* 456:93–97.

Keith, L. B. 1963. *Wildlife's ten-year cycle.* Madison: Univ. of Wisconsin Press.

———. 1983. Role of food in hare populations. *Oikos* 40:385–395.

———. 1987. Dynamics of snowshoe hare populations. *Curr. Mammal.* 2:119–195.

Kemp, T. S. 2006. The origin of mammalian endothermy: A paradigm for the evolution of complex biological structure. *Zool. J. Linnean Soc.* 147:473–488.

Kenagy, G. J. 1972. Saltbush leaves: Excision of hypersaline tissue by a kangaroo rat. *Science* 178:1094–1096.

Kerley, G. I. H., and W. G. Whitford. 1994. Desert-dwelling small mammals as granivores: Intercontinental variations. *Aust. J. Zool.* 42:543–555.

Kido, H., and D. Uemura. 2004. *Blarina* toxin, a mammalian lethal venom from the short-tailed shrew *Blarina brevicauda*: Isolation and characterization. *Proc. Natl. Acad. Sci. USA* 101:7542–7547.

Kimura, K. A., and T. A. Uchida. 1983. Ultrastructural observations of delayed implantation in the Japanese long-fingered bat, *Miniopterus schreibersii fulginosis. J. Reprod. Fertil.* 69:187–193.

King, C. 1983. *Mustela erminea. Mammal. Spec.* 195:1–8.

King, C. M. 1989. The advantages and disadvantages of small size to weasels, *Mustela* species. In *Carnivore behavior, ecology, and evolution,* ed. J. L. Gittleman, 302–334. Ithaca, NY: Comstock.

———. 1990. *The natural history of weasels and stoats.* Ithaca, NY: Comstock.

King, J. A., D. Maas, and R. G. Weisman. 1964. Geographic variation in nest size among species of *Peromyscus. Evolution* 18:230–234.

Kingdon, J. 1974. *East African mammals,* vol. 2. New York: Academic Press.

———. 1997a. *East African mammals: Large mammals,* vol. 3B. Chicago: Univ. of Chicago Press.

———. 1997b. *The Kingdon field guide to African mammals.* New York: Academic Press.

Kita, M., Y. Okumura, S. D. Ohdachi, Y. Oba, M. Yoshikuni, Y. Nakamura, H. Kido, and D. Uemura. 2005. Purification and characterization of blarinasin, a new tissue kallikrein-like protease from the short-tailed shrew *Blarina brevicauda*: Comparative studies with *Blarina* toxin. *Biol. Chem.* 386:177–182.

Kleiber, M. 1932. Body size and metabolism. *Hilgardia* 6:315–353.

———. 1961. *The fire of life*. New York: Wiley.

Kleiman, D. G. 1977. Monogamy in mammals. *Q. Rev. Biol.* 52:39–69.

Kliman, R. M., and G. R. Lynch. 1992. Evidence for genetic variation in the occurrence of the photoresponse of the Djungarian hamster, *Phodopus sungorus*. *J. Biol. Rhythms* 7:161–173.

Klir, J. J., and J. E. Heath. 1992. An infrared thermographic study of surface temperature in relation to external thermal stress in three species of foxes: The red fox (*Vulpes vulpes*), arctic fox (*Alopex lagopus*), and kit fox (*Vulpes macrotis*). *Physiol. Zool.* 65:1011–1021.

Koehler, C. E., and P. R. K. Richardson. 1990. *Proteles cristatus. Mamm. Species* 363:1–6.

Körtner, G., and F. Geiser. 1998. Ecology of natural hibernation in the marsupial mountain pygmy-possum (*Burramys parvus*). *Oecologia* 113:170–178.

Krakauer, E., P. Lemelin, and D. Schmitt. 2002. Hand and body position during locomotion behavior in the aye-aye (*Daubentonia madagascariensis*). *Am. J. Primatol.* 57:105–118.

Krebs, C. J. 1996. Population cycles revisited. *J. Mammal.* 77:8–24.

Krebs, C. J., S. Boutin, R. Boonstra, A. R. E. Sinclair, J. N. M. Smith, M. R. T. Dale, and R. Turkington. 1995. Impact of food and predation on the snowshoe hare cycle. *Science* 269:1112–1115.

Krebs, C. J., and J. H. Myers. 1974. Population cycles in small mammals. *Adv. Ecol. Res.* 8:367–399.

Kruuk, H., and W. A. Sands. 1972. The aardwolf (*Proteles cristatus* Sparrman) 1783 as a predator of termites. *E. Afr. Wildl. J.* 10:211–227.

Kuchel, L. 2003 The energetics and patterns of torpor in free-ranging echidnas (*Tachyglossus aculeatus*) from a warm temperature climate. Ph.D. diss., Univ. of Queensland, Australia.

Kunz, T. H., and M. B. Fenton. 2003. *Bat ecology.* Chicago: Univ. of Chicago Press.

Kurta, A. 1995. *Mammals of the Great Lakes region.* Ann Arbor: Univ. of Michigan Press.

———. 2008. *Bats of Michigan.* Indiana State Univ. Center for North American Bat Research and Conservation pub. no. 2. Terre Haute: Indiana State Univ. Center for North American Bat Research and Conservation.

Kurta, A., K. A. Johnson, and T. H. Kunz. 1987. Oxygen consumption and body temperature of female little brown bats (*Myotis lucifugus*) under simulated roost conditions. *Physiol. Zool.* 60:386–397.

Lacey, E. A., and P. W. Sherman. 1997. Cooperative breeding in naked mole-rats: Implications for vertebrate and invertebrate sociality. In *Cooperative breeding in mammals,* ed. N. G. Solomon and J. A. French, 267–301. Cambridge: Cambridge Univ. Press.

Lang, H., and J. P. Chapin. 1917. Notes on the distribution and ecology of central African Chiroptera. *Bull. Am. Mus. Nat. Hist.* 37:479–563.

Larivière, S. 1999. *Mustela vison. Mamm. Species* 608:1–9.

———. 2002. *Vulpes zerda. Mamm. Species* 714:1–5.

Larivière, S., and P. J. Seddon. 2001. *Vulpes rueppelli. Mamm. Species* 678:1–5.

Law, B. S. 1992. Physiological factors affecting pollen use by the Queensland blossom bat (*Syconycteris australis*). *Funct. Ecol.* 6:257–264.

———. 1993. Roosting and foraging ecology of the Queensland blossom-bat (*Syconycteris australis*) in northeastern New South Wales: Flexibility in response to seasonal variation. *Wildl. Res.* 20:419–431.

Laws, R. M., A. Baird, and M. M. Bryden. 2003. Breeding season and embryonic diapause in crabeaster seals (*Lobodon carcinophagus*). *Reproduction* 126:365–370.

Layne, J. N. 1969. Nest-building behavior in three species of deer mice, *Peromyscus. Behaviour* 35:288–303.

Lee, A. K., and A. Cockburn. 1985. *Evolutionary ecology of marsupials.* New York: Cambridge: Cambridge Univ. Press.

Leiner, N. O., E. Z. F. Setz, and W. R. Silva. 2008. Semelparity and factors affecting the reproductive activity of the Brazilian slender opossum (*Marmosops paulensis*) in southeastern Brazil. *J. Mammal.* 89:153–158.

Lessa, E. P., and J. A. Cook. 1998. The molecular phylogenetics of tuco-tucos (genus *Ctenomys,* Rodentia: Octodontidae) suggests an early burst of speciation. *Mol. Phylogenet. Evol.* 9:88–99.

Lewis, E. R., P. M. Narins, J. U. M. Jarvis, G. Bronner, and M. J. Mason. 2006. Preliminary evidence for the use of microseismic cues for navigation by the Namib golden mole. *J. Acoust. Soc. Am.* 119:1260–1268.

Lidicker, W. Z. Jr. 1988. Solving the enigma of microtine "cycles." *J. Mammal.* 69:225–235.

Lovegrove, B. G. 2000. Daily heterothermy in mammals: Coping with unpredictable environments. In *Life in the cold,* ed. G. Heldmaier and M. Klingenspor, 29–40. Eleventh International Hibernation Symposium, 13–18 August 2000, Jungholz, Austria. Berlin: Springer-Verlag.

———. 2005. Seasonal thermoregulatory responses in mammals. *J. Comp. Physiol. B* 175:231–247.

Lurz, P. W. W., and A. B. South. 1998. Cached fungi in non-native conifer forest and their importance for red squirrels (*Sciurus vulgaris* L.). *J. Zool. Lond.* 246:468–471.

MacArthur, R. A., and M. Aleksiuk. 1979. Seasonal microenvironments of the muskrat (*Ondatra zibethicus*) in a northern marsh. *J. Mammal.* 60:146–154.

Macdonald, D., ed. 2006. *The encyclopedia of mammals.* New York: Facts on File.

MacLulich, D. A. 1937. *Fluctuations in the numbers of the varying hare.* Univ. of Toronto Studies, Biology Series, no. 43. Toronto: Univ. of Toronto Press.

Madison, D. M. 1984. Group nesting and its ecological and evolutionary significance in overwintering microtine rodents. In *Winter ecology of small mammals,* ed. J. F. Merritt, 267–274. Carnegie Museum of Natural History spec. pub. no. 10. Pittsburgh: Carnegie Museum of Natural History.

Mangan, S. A., and G. H. Adler. 2002. Seasonal dispersal of arbuscular mycorrhizal fungi by spiny rats in a neotropical forest. *Oecologia* 131:587–597.

Manger, P. R., and J. D. Pettigrew. 1995. Electroreception and the feeding behaviour of platypus (*Ornithorhynchus anatinus*: Monotremata: Mammalia). *Phil. Trans. R. Soc. Lond. Ser. B* 347:359–381.

Manser, M. B. 2001. The acoustic structure of suricates' alarm calls varies with predator type and the level of response urgency. *Proc. R. Soc. Lond. Ser. B* 268:2315–2324.

Mansergh, I. M., and L. S. Broome. 1994. *The mountain pygmy-possum of the Australian Alps.* Sydney: New South Wales Univ. Press.

Marchand, P. J. 1996. *Life in the cold: An introduction to winter ecology.* 3d ed. Hanover, NH: Univ. Press of New England.

Mares, M., and M. L. Rosenzweig. 1978. Granivory in North and South American deserts: Rodents, birds, and ants. *Ecology* 59:234–241.

Marsden, W. 1964. *The lemming year.* London: Chatto and Windus.

Marshall, L. G. 1978. *Chironectes minimus. Mamm. Species* 109:1–6.

Martin, I. G. 1983. Daily activity of short-tailed shrew (*Blarina brevicauda*) in simulated natural conditions. *Am. Midl. Nat.* 109:136–144.

Mason, M. J. 2003. Morphology of the middle ear of golden moles (Chrysochloridae). *J. Zool. Lond.* 260:391–403.

Mason, M. J., and P. M. Narins. 2002. Seismic sensitivity in the desert gold mole (*Eremitalpa granti*): A review. *J. Comp. Psychol.* 116:158–163.

Mauck, B., K. Bilgmann, D. D. Jones, U. Eysel, and G. Dehnhardt. 2003. Thermal windows on the trunk of hauled-out seals: Hot spots for thermoregulatory evaporation? *J. Exp. Biol.* 206:1727–1738.

Maynard, C. H. 1889. Singular effects produced by the bite of a short-tailed shrew, *Blarina brevicauda. Contrib. Sci.* 1:57–59.

Mayr, E. 1956. Geographical character gradients and climatic adaptation. *Evolution* 10:105–108.

———. 1970. *Populations, species, and evolution.* Cambridge, MA: Belknap Press of Harvard Univ. Press.

McAllan, B. 2003. Timing of reproduction in carnivorous marsupials. In *Predators with pouches,* ed. M. Jones, C. R. Dickman, and M. Archer, 147–168. Collingwood, Australia: CSIRO.

McCracken, G. F. 2003. Estimates of population sizes in summer colonies of Brazilian free-tailed bats (*Tadarida brasiliensis*). In *Monitoring trends in bat populations of the U.S. and territories: Problems and prospects,* ed. T. J. O'Shea and M. A. Bogan, 21–30. U.S. Geological Survey, Biological Resources Discipline, Information and Technology Report, USGS/BRD/ITR-2003-003:21-30. Fort Collins, CO: U.S. Geological Survey.

McGrew, J. C. 1979. *Vulpes macrotis. Mamm. Species* 123:1–6.

McKenna, M. C., and S. K. Bell. 1997. *Classification of mammals above the species level.* New York: Columbia Univ. Press.

McNab, B. K. 1971. On the ecological significance of Bergmann's rule. *Ecology* 52:845–854.

———. 1989. Basal rate of metabolism, body size, and food habits in the order Carnivora. In *Carnivore behavior, ecology, and evolution,* ed. J. L. Gittleman, 335–354. Ithaca, NY: Comstock.

———. 2002. *The physiological ecology of vertebrates: A view from energetics.* Ithaca, NY: Comstock.

McNab, B. K., and P. Morrison. 1963. Body temperature and metabolism in subspecies of *Peromyscus* from arid and mesic environments. *Ecol. Monogr.* 33:63–82.

Mead, R. A. 1989. The physiology and evolution of delayed implantation in carnivores. In *Carnivore behavior, ecology, and evolution,* ed. G. L. Gittleman, 437–464. Ithaca, NY: Comstock.

Merritt, J. F. 1978. *Peromyscus californicus. Mamm. Species* 85:1–6.

———. 1986. Winter survival adaptations of the short-tailed shrew (*Blarina brevicauda*) in an Appalachian montane forest. *J. Mammal.* 67:450–464.

———. 1995. Seasonal thermogenesis and changes in body mass of masked shrews, *Sorex cinereus. J. Mammal.* 76:1020–1035.

Merritt, J. F., and J. M. Merritt. 1978. Population ecology and energy relationships of *Clethrionomys gapperi* in a Colorado subalpine forest. *J. Mammal.* 59:576–598.

Merritt, J. F., and D. A. Zegers. 2002. Maximizing survivorship in cold: Thermogenic profiles of non-hibernating mammals. *Acta Theriol.* 47:221–234.

Merritt, J. F., D. A. Zegers, and L. R. Rose. 2001. Seasonal thermogenesis of southern flying squirrels (*Glaucomys volans*). *J. Mammal.* 82:51–64.

Milner, J. M., and S. Harris. 1999. Activity patterns and feeding behaviour of the tree hyrax, *Dendrohyrax arboreus,* in the Parc National des Volcans, Rwanda. *Afr. J. Ecol.* 37:267–280.

Mitchell, D., S. K. Maloney, H. P. Laburn, M. H. Knight, G. Kuhnen, and C. Jessen. 1997. Activity, blood temperature, and brain temperature of free-ranging springbok. *J. Comp. Physiol. B* 167:335–343.

Montgomery, G. G. 1978. *The ecology of arboreal folivores.* Washington, DC: Smithsonian Institution Press.

Morgan, K. H., and M. V. Price. 1992. Foraging in heteromyid rodents: The energy of scratch-digging. *Ecology* 73:2260–2272.

Morrison, P. 1966. Insulative flexibility in the guanaco. *J. Mammal.* 47:18–23.

Morton, S. R. 1980. Field and laboratory studies of water metabolism in *Sminthopsis crassicaudata* (Marsupialia, Dasyuridae). *Aust. J. Zool.* 28:213–227.

———. 1985. Granivory in arid regions: Comparison of Australia and North and South America. *Ecology* 66:1859–1866.

Muchhala, N. 2006. Nectar bat stows huge tongue in its rib cage. *Nature* 444:701.

Muchhala, N., and P. Jarrin-V. 2002. Flower visitation by bats in cloud forests of western Ecuador. *Biotropica* 34:387–395.

Muchhala, N., P. Mena, and L. Albuja V. 2005. A new species of *Anoura* (Chiroptera: Phyllostomidae) from the Ecuadorian Andes. *J. Mammal.* 86:457–461.

Munshi-South, J., L. H. Emmons, and H. Bernard. 2007. Behavioral monogamy and fruit availability in the large treeshrew (*Tupaia tana*) in Sabah, Malaysia. *J. Mammal.* 88:1427–1438.

Murray, B. D., and C. R. Dickman. 1994. Granivory and microhabitat use in Australian desert rodents: Are seeds important? *Oecologia* 99:216–225.

Muul, I. 1968. Behavioral and physiological influences on the distribution of the flying squirrel, *Glaucomys volans. Misc. Publ. Mus. Zool. Univ. Mich.* 124:1–66.

Nagel, A. 1977. Torpor in the European white-toothed shrews. *Experientia* 33:1455–1458.

Narins, P. M., E. R. Lewis, J. U. M. Jarvis, and J. O'Riain. 1997. The use of seismic signals by fossorial southern African mammals: A neuroethological gold mine. *Brain Res. Bull.* 44:641–646.

Narins, P. M., O. J. Reichman, J. U. M. Jarvis, and E. R. Lewis. 1992. Seismic signal transmission between burrows of the Cape mole-rat, *Georychus capensis. J. Comp. Physiol. A* 170:13–21.

Nash, L. T. 1986. Dietary, behavioral, and morphological aspects of gummivory in primates. *Yearbook Phys. Anthropol.* 29:113–137.

Nelson, J. F., and R. M. Chew. 1977. Factors affecting seed reserves in the soil of a Mojave Desert ecosystem, Rock Valley, Nye County, Nevada. *Am. Midl. Nat.* 97:300–320.

Neuweiler, G. 2000. *The biology of bats.* Oxford: Oxford Univ. Press.

Nicol, S. C., and N. A. Andersen. 1993. The physiology of hibernation in an egg-laying mammal, the echidna. In *Life in the cold: Ecological, physiological, and molecular mechanisms,* ed. C. Carey, G. L. Florant, B. A. Wunder, and B. Horwitz, 56–64. Boulder, CO: Westview Press.

Nicoll, M. E., and P. A. Racey. 1985. Follicular development, ovulation, fertilization, and fetal development in tenrecs (*Tenrec ecaudatus*). *J. Reprod. Fertil.* 74:47–55.

Niethammer, G. 1970. Beobachtungen am Pyrenaen-Desman, *Galemys pyrenaica. Bonner Zool. Beitr.* 21:157–182.

Ninomiya, H. 2000. The vascular bed in the rabbit ear: Microangiography and scanning electron microscopy of vascular corrosion casts. *Anat. Histol. Embryol.* 29:301–305.

Noll-Banholzer, U. 1979. Body temperature, oxygen consumption, evaporative water loss, and heart rate in the fennec. *Comp. Biochem. Physiol.* 62A:585–592.

Norberg, U. M. 1990. *Vertebrate flight: Mechanisms, physiology, morphology, ecology, and evolution.* New York: Springer-Verlag.

———. 1994. Wing design, flight performance, and habitat use in bats. In *Ecological morphology: Integrative organismal biology,* ed. P. C. Wainwright and S. M. Reilly, 205–239. Chicago: Univ. of Chicago Press.

Norberg, U. M., and M. B. Fenton. 1988. Carnivorous bats? *Biol. J. Linnean Soc.* 33:383–394.

Nowak, R. M. 1999. *Walker's mammals of the world.* 6th ed., vols. 1 and 2. Baltimore: Johns Hopkins Univ. Press.

Oakwood, M., A. J. Bradley, and A. Cockburn. 2001. Semelparity in a large marsupial. *Proc. R. Soc. Lond. B Biol. Sci.* 268:407–411.

Oehler, C. 1944. Notes on the temperament of the New York weasel. *J. Mammal.* 25:198.

O'Riain, M. J., N. C. Bennett, P. N. M. Brotherton, G. McIlrath, and T. H. Clutton-Brock. 2000. Reproductive suppression and inbreeding avoidance in wild populations of cooperatively breeding meerkats (*Suricata suricatta*). *Behav. Ecol. Sociobiol.* 48:471–477.

O'Riain, M. J., and S. Braude. 2001. Inbreeding versus outbreeding in captive and wild populations of naked mole-rats. In *Dispersal,* ed. J. Clobert, E. Dhondt, and J. Nichols, 143–154. Oxford: Oxford Univ. Press.

Øritsland, N. A., J. W. Lentfer, and K. Ronald. 1974. Radiative surface temperature of the polar bear. *J. Mammal* 55:459–461.

Osborn, D. J., and I. Helmy. 1980. *The contemporary land mammals of Egypt (including Sinai).* Chicago: Field Museum of Natural History.

Packard, G. C., and T. J. Boardman. 1988. The misuse of ratios, indices, and percentages in ecophysiological research. *Physiol. Zool.* 61:1–9.

Padykula, H. A., and J. M. Taylor. 1982. Marsupial placentation and its evolutionary significance. *J. Reprod. Fertil. Suppl.* 31:95–104.

Patterson, B. D., M. R. Willig, and R. D. Stevens. 2003. Trophic strategies, niche partitioning, and patterns of ecological organization. In *Bat ecology,* ed. T. H. Kunz and M. B. Fenton, 536–579. Chicago: Univ. of Chicago Press.

Phillips, P. K., and J. E. Heath. 1992. Heat exchange by the pinna of the African elephant (*Loxodonta africana*). *Comp. Biochem. Physiol.* 101A:693–699.

———. 2001. An infrared thermographic study of surface temperature in the eu-thermic woodchuck (*Marmota monax*). *Comp. Biochem. Physiol.* 129A:557–562.

Pitelka, F. A. 1958. Some aspects of population structure in the short-term cycle of the brown lemming in northern Alaska. *Cold Spring Harb. Symp. Quant. Biol.* 22:237–251.

Poglayen-Neuwall, I., and D. E. Toweill. 1988. *Bassariscus astutus. Mamm. Species* 327:1–8.

Popa-Lisseanu, A. G., A. Delgado-Huertas, M. G. Forero, A. Rodríguez, R. Arlet-taz, and C. Ibáñez. 2007. Bats' conquest of a formidable foraging niche: The myriads of nocturnally migrating songbirds. *PLoS ONE* 2(2):e205.

Post, D. M., O. J. Reichman, and D. E. Wooster. 1993. Characteristics and signi-ficance of the caches of eastern woodrats (*Neotoma floridana*). *J. Mammal.* 74:688–692.

Post, D. M., M. V. Snyder, E. J. Finck, and D. K. Saunders. 2006. Caching as a strat-egy for surviving periods of resource scarcity: A comparative study of two spe-cies of *Neotoma. Funct. Ecol.* 20:717–722.

Powell, R. A. 1993. *The fisher: Life history, ecology, and behavior.* 2d ed. Min-neapolis: Univ. of Minnesota Press.

Prestrud, P. 1991. Adaptations by the arctic fox (*Alopex lagopus*) to the polar win-ter. *Arctic* 44:132–138.

Price, M. V., N. M. Waser, and S. McDonald. 2000. Seed caching by heteromyid rodents from two communities: Implications for coexistence. *J. Mammal.* 81:97–106.

Proske, U., and E. Gregory. 2003. Electrolocation in the platypus—some specula-tions. *Comp. Biochem. Physiol. A* 136:821–825

Proske, U., J. E. Gregory, and A. Iggo. 1998. Sensory receptors in monotremes. *Phil. Trans. R. Soc. Lond. Ser. B Biol. Sci.* 353:1187–1198.

Puig, S., M. I. Rosi, M. I. Cona, V. G. Roig, and S. A. Monge. 1999. Diet of a pied-mont population of *Ctenomys mendocinus* (Rodentia, Ctenomyidae): Seasonal patterns and variations according to sex and relative age. *Acta Theriol.* 44:15–27.

Quinn, A., and D. E. Wilson. 2004. *Daubentonia madagascariensis. Mamm. Spe-cies* 740:1–6.

Racey, P. A. 1982. Ecology of bat reproduction. In *Ecology of bats,* ed. T. H. Kunz, 57–104. New York: Plenum.

Rado, R., N. Levi, H. Hauser, J. Witcher, N. Alder, N. Intrator, Z. Wollberg, and J. Terkel. 1987. Seismic signaling as a means of communication in a subter-ranean mammal. *Anim. Behav.* 35:1249–1251.

Randall, J. A. 1993. Behavioural adaptations of desert rodents (Heteromyidae). *Anim. Behav.* 45:263–287.

Ransome, R. 1990. *The natural history of hibernating bats.* London: Christopher Helm.

Rathbun, G. B. 1979. The social structure and ecology of elephant-shrews. *Adv. Ethol.* 20:1–75.

Rathbun, G. B., P. Beaman, and E. Maliniak. 1981. Captive, husbandry and breeding of rufous elephant-shrews. *Int. Zoo Yearbook* 21:176–184.

Rathbun, G. B., and C. D. Rathbun. 2006. Social structure of the bushveld sengi (*Elephantulus intufi*) in Namibia and the evolution of monogamy in the Macroscelidea. *J. Zool.* 269:391–399.

Rathbun, G. B., and K. Redford. 1981. Pedal scent-marking in the rufous elephant-shrew *Elephantulus rufescens. J. Mammal* 62:635–637.

Rauch, J. C., and J. S. Hayward. 1969. Topography and vascularization of brown fat in a hibernator (little brown bat, *Myotis lucifugus*). *Can. J. Zool.* 47:1315–1323.

Reichard, U. H., and C. Boesch, eds. 2003. *Monogamy: Mating strategies and partnerships in birds, humans and other mammals.* Cambridge: Cambridge Univ. Press.

Reichman, O. J., and C. Rebar. 1985. Seed preferences by desert rodents based on levels of mouldiness. *Anim. Behav.* 33:726–729.

Reid, F. A. 2006. *A field guide to mammals of North America north of Mexico.* 4th ed. Peterson Field Guide Series. Boston: Houghton Mifflin.

Reiss, K. Z. 1997. Myology of the feeding apparatus of myrmecophagid anteaters (Xenarthra: Myrmecophagidae). *J. Mammal. Evol.* 4:87–117.

Renfree, M. B., E. M. Russell, and R. D. Wooller. 1984. Reproduction and life history of the honey possum, *Tarsipes rostratus.* In *Possums and gliders,* ed. A. P. Smith and I. D. Hume, 427–437. Sydney, Australia: Surrey Beatty.

Rensch, B. 1936. Studien über klimatische paarallelitat der merkmalsauspragung bei vogeln und saugern. *Arch. Naturgeschichte N.F.* 5:17–363.

———. 1938. Some problems of geographical variation and species-formation. *Proc. Linnean Soc. Lond.* 150:275–285.

Richard, P. B. 1982. La sensibilite tactile de contact chez le desman (*Galemys pyrenaicus*). *Biol. Behav.* 7:325–336.

Richards, G. C. 1990. Rainforest bat conservation: Unique problems in a unique environment. *Aust. J. Zool.* 26:44–46.

Richardson, E. G. 1977. The biology and evolution of the reproductive cycle of *Miniopterus schreibersii* and *M. australis* (Chiroptera: Vespertilionidae). *J. Zool.* 183:353–375.

Rismiller, P. D. 1999. *The echidna: Australia's enigma.* Hong Kong: Hugh Lauter Levin.

Rismiller, P. D., and M. W. McKelvey. 2000. Frequency of breeding and recruitment in the short-beaked echidna, *Tachyglossus aculeatus. J. Mammal.* 81:1–17.

Rosenzweig, M. L. 1968. The strategy of body size in mammalian carnivores. *Am. Midl. Nat.* 80:299–315.

Russell, E. M. 1982. Patterns of parental care and parental investment in marsupials. *Biol. Rev. Camb. Phil. Soc.* 57:423–486.

Ryan, M. J., and M. D. Tuttle. 1983. The ability of the frog-eating bat to discriminate among novel and potentially poisonous frog species using acoustic cues. *Anim. Behav.* 31:827–833.

Sandell, M. 1984. To have or not to have delayed implantation: The example of the weasel and the stoat. *Oikos* 42:123–126.

Sargis, E. 2004. New views on tree shrews: The role of tupaiids in primate supraordinal relationships. *Evol. Anthropol.* 13:56–66.

Scantlebury, M., M. K. Oosthuizen, J. R. Speakman, C. R. Jackson, and N. C. Bennett. 2005. Seasonal energetics of the Hottentot golden mole at 1500 altitude. *Physiol. Behav.* 84:739–745.

Scheich, H., G. Langner, C. Tidemann, R. B. Coles, and A. Guppy. 1986. Electroreception and electrolocation in platypus. *Nature* 319:401–402.

Schlitter, D. A. 2005. Order Macroscelidea. In *Mammal species of the world: A taxonomic and geographic reference*, 3d ed., vol. 1, ed. D. E. Wilson and D. M. Reeder, 82–85. Baltimore: Johns Hopkins Univ. Press.

———. 1997. *Animal physiology: Adaptation and environment.* 5th ed. New York: Cambridge Univ. Press.

Schmidt-Nielsen, K., and R. O'Dell. 1961. Structure and concentrating mechanism in the mammalian kidney. *Am. J. Physiol.* 200:1119–1124.

Schnitzler, H. U. 1972. Control of Doppler shift compensation in the greater horseshoe bat, *Rhinolophus ferrumequinum. J. Comp. Physiol.* 82:79–92.

Schnitzler, H. U., E. K. V. Kalko, I. Kaipf, and A. D. Grinnell. 1994. Fishing and echolocation behavior of the greater bulldog bat, *Noctilio leporinus,* in the field. *Behav. Ecol. Sociobiol.* 35:327–345.

Scholander, P. F. 1955. Evolution of climatic adaptation in homeotherms. *Evolution* 9:15–26.

———. 1957. The wonderful net. *Sci. Am.* 196:97–107.

Scholander, P. F., V. Waters, R. Hock, and L. Irving. 1950. Body insulation of some arctic and tropical mammals and birds. *Biol. Bull.* 99:225–235.

Schreber, J. C. D. 1774. *Die Säugthiere in Abbildungen nach der Natur, mit Beschreibungen.* Erlangen, Germany: Erster Theil.

Schuller, G., and G. Pollak. 1979. Disproportionate frequency representation in the inferior colliculus of Doppler compensating greater horseshoe bats: Evidence for an acoustic fovea. *J. Comp. Physiol.* 132:47–54.

Sealander, J. A. 1952. The relationship of nest protection and huddling to survival of *Peromyscus* at low temperature. *Ecology* 33:63–71.

Seton, E. T. 1911. *The arctic prairies.* New York: Scribners.

————. 1929. *Lives of game animals,* vol. II, pt. II: *Bears, coons, badgers, skunks, and weasels.* New York: Doubleday, Doran.

Sharpe, D. J., and R. L. Goldingay. 2007. Home range of the Australian squirrel glider, *Petaurus norfolcensis* (Diprodontia). *J. Mammal* 88:1515–1522.

Sheriff, M. J., L. Kuchel, S. Boutin, and M. M. Humphries. 2009. Seasonal metabolic acclimatization in a northern population of free-ranging snowshoe hares, *Lepus americanus. J. Mammal.* 90:761–767

Sherman, P. W., S. Braude, and J. U. M. Jarvis. 1999. Litter sizes and mammary numbers of naked mole-rats: Breaking the one-half rule. *J. Mammal.* 80:720–733.

Sherman, P. W., J. U. M. Jarvis, and R. D. Alexander, eds. 1991. *The biology of the naked mole-rat.* Princeton, NJ: Princeton Univ. Press.

Sherrod, S. K., T. R. Seastedt, and M. D. Walker. 2005. Northern pocket gopher (*Thomomys talpoides*) control of alpine plant community structure. *Arct. Antarct. Alpine Res.* 37:585–590.

Silva, R. B., E. M. Vieira, and P. Izar. 2008. Social monogamy and biparental care of the neotropical southern bamboo rat (*Kannabateomys amblyonyx*). *J. Mammal.* 89:1464–1472.

Simmons, N. B. 2005. Order Chiroptera. In *Mammal species of the world: A taxonomic and geographic reference,* 3d ed., vol. 1, ed. D. E. Wilson and D. M. Reeder, 312–529. Baltimore: Johns Hopkins Univ. Press.

Simmons, N. B., K. L. Seymour, J. Habersetzer, and G. F. Gunnell. 2008. Primitive early Eocene bat from Wyoming and the evolution of flight and echolocation. *Nature* 451:818–821.

Simms, D. A. 1979. North American weasels: Resource utilization and distribution. *Can. J. Zool.* 57:504–520.

Sinclair, A. R. E., and J. M. Gosline. 1997. Solar activity and mammal cycles in the northern hemisphere. *Am. Nat.* 149:776–784.

Sinclair, A. R. E., J. M. Gosline, G. Holdsworth, C. J. Krebs, S. Boutin, J. N. M. Smith, R. Boonstra, and M. Dale. 1993. Can the solar cycle and climate synchronize the snowshoe hare cycle in Canada? *Am. Nat.* 141:173–198.

Sliwa, A. 1994. Diet and feeding behaviour of the black-footed cat (*Felis nigripes* Burchell, 1824) in the Kimberley Region, South Africa. *Zool. Garten N.F.* 64: 83–96.

Smith, A. P., and L. Broome. 1992. The effects of season, sex and habitat on the diet of the mountain pygmy possum (*Burramys parvus*). *Wildl. Res.* 19:755–768.

Sommer, S. 1997. Monogamy in *Hypogeomys antimena,* an endemic rodent of the deciduous dry forest in western Madagascar. *J. Zool. Lond.* 241/242:301–314.

Spallanzani, L. 1932. *Opere di Lazzaro Spallanzani,* vol. 3. Milan: Ulrico Heopli.

Sparti, A. 1992. Thermogenic capacity of shrews (Mammalia, Soricidae) and its relationship with basal rate of metabolism. *Physiol. Zool.* 65:77–96.

Spencer, D. A. 1958. Biological and control aspects. In *The Oregon meadow mouse irruption of 1957–1958*, 15–25. Corvallis, OR: Federal Cooperative Extension Service, Oregon State College.

Sperber, I. 1944. Studies on the mammalian kidney. *Zool. Bidrag Fran Uppsala* 22:249–430.

Springer, M. S., M. J. Stanhope, O. Madsen, and W. W. DeJong. 2004. Molecules consolidate the placental mammal tree. *Trends Ecol. Evol.* 19:430–438.

Stafford, B. J. 2005. Order Dermoptera. In *Mammal species of the world: A taxonomic and geographic reference,* 3d ed., vol. 1, ed. D. E. Wilson and D. M. Reeder, 110. Baltimore: Johns Hopkins Univ. Press.

Stapp, P., P. J. Pekins, and W. W. Mautz. 1991. Winter energy expenditure and the distribution of southern flying squirrels. *Can. J. Zool.* 69:2548–2555.

Stearns, S. C. 1992. *The evolution of life histories.* New York: Oxford Univ. Press.

Steele, M. A., and J. L. Koprowski. 2001. *North American tree squirrels.* Washington, DC: Smithsonian Institution Press.

Steele, M. A., S. Manierre, T. Genna, T. A. Contreras, P. D. Smallwood, and M. E. Pereira. 2006. The innate basis of food-hoarding decisions in grey squirrels: Evidence for behavioural adaptations to the oaks. *Anim. Behav.* 71:155–160.

Stenseth, N. C., and R. A. Ims. 1993. *The biology of lemmings.* London: Academic Press.

Stephenson, A. B. 1969. Temperature within a beaver lodge in winter. *J. Mammal.* 50:134–136.

Stephenson, P. J., and P. A. Racey. 1994. Seasonal variation in resting metabolic rate and body temperature of streaked tenrecs, *Hemicentetes nigriceps* and *H. semispinosus* (Insectivora, Tenrecidae). *J. Zool.* 232:285–294.

Steudel, K., W. P. Porter, and D. Sher. 1994. The biophysics of Bergmann's rule—a comparison of the effects of pelage and body-size variation on metabolic rate. *Can. J. Zool.* 72:70–77.

Stevens, C. E., and I. D. Hume. 1996. *Comparative physiology of the vertebrate digestive system.* 2d ed. New York: Cambridge Univ. Press.

Stevenson, R. D. 1986. Allen's rule in North American rabbits (*Sylvilagus*) and hares (*Lepus*) is an exception, not a rule. *J. Mammal.* 67:312–316.

Stoddart, D. M., ed. 1979. *Ecology of small mammals.* London: Chapman and Hall.

Stoner, C. J., O. R. P. Bininda-Emonds, and T. Caro. 2003. The adaptive significance of coloration in lagomorphs. *Biol. J. Linnean Soc.* 79:309–328.

Storey, K. B., and J. M. Storey. 1988. Freeze tolerance in animals. *Physiol. Rev.* 68:27–84.

Svendsen, G. E. 1982. Weasels, *Mustela* species. In *Wild mammals of North America: Biology, management, and economics,* ed. J. A. Chapman and G. A. Feldhamer, 613–628. Baltimore: Johns Hopkins Univ. Press.

Tattersall, I. 1982. *The primates of Madagascar.* New York: Columbia Univ. Press.

Taylor, C. R. 1972. The desert gazelle: A paradox resolved. In *Comparative physiology of desert animals,* ed. G. M. O. Maloiy, 215–227. Symp. Zool. Soc. Lond., no. 31. London: Zoological Society of London.

Taylor, C. R., and C. P. Lyman. 1972. Heat storage in running antelopes: Independence of brain and body temperatures. *Am. J. Physiol.* 222:114–117.

Taylor, J. R. E. 1998. Evolution of energetic strategies of shrews. In *Evolution of shrews,* ed. J. M. Wójcik and M. Wolsan, 309–346. Białowieza, Poland: Mammal Research Institute, Polish Academy of Sciences.

Thabah, A., G. Li, Y. Wang, B. Liang, K. Hu, S. Zhang, and G. Jones. 2007. Diet, echolocation calls and phylogenetic affinities of the great evening bat *Ia io* (Vespertilionidae): Another carnivorous bat. *J. Mammal.* 88:728–735

Thomas, D. W., M. Dorais, and J.-M. Bergeron 1990. Winter energy budgets and costs of arousals for hibernating little brown bats, *Myotis lucifugus. J. Mammal.* 71:475–479.

Tomasi, T. E. 1978. Function of venom in the short-tailed shrew, *Blarina brevicauda. J. Mammal.* 59:852–854.

Tracy, R. L., and G. E. Walsberg. 2001. Intraspecific variation in water loss in a desert rodent, *Dipodomys merriami. Ecology* 82:1130–1137.

Trapp, G. R. 1972. Some anatomical and behavioral adaptations of ringtails, *Bassariscus astutus. J. Mammal.* 53:549–557.

Trappe, J. M., and C. Maser. 1976. Germination of spores of *Glomus macrocarpus* (Endogonaceae) after passage through a rodent digestive tract. *Mycologia* 68:433–436.

Trayhurn, P., J. H. Beattie, and D. V. Rayner. 2000. Leptin-signals and secretions from white adipose tissue. In *Life in the cold,* ed. G. Heldmaier and M. Klingenspor, 459–469. Eleventh International Hibernation Symposium, 13–18 August 2000, Jungholz, Austria. Berlin: Springer-Verlag,.

Turner, M. I. A., and R. M. Watson. 1965. An introductory study on the ecology of hyrax (*Dendrohyrax brucei* and *Procavia johnstoni*) in the Serengeti National Park. *East Afr. Wildl. J.* 3:49–60.

Tuttle, M. D., and M. J. Ryan. 1981. Bat predation and the evolution of frog vocalizations in the neotropics. *Science* 214:677–678.

Tyndale-Biscoe, C. H. 2005. *Life of marsupials.* Collingwood, Australia: CSIRO.

Underwood, L. S., and P. Reynolds. 1980. Photoperiod and fur lengths in the arctic fox (*Alopex lagopus*). *Int. J. Biometeorol.* 24:39–48.

Van Dyck, S., and R. Strahan, eds. 2008. *The mammals of Australia.* 3d ed. Sydney, Australia: New Holland.

Vander Wall, S. B. 1990. *Food hoarding in animals.* Chicago: Univ. of Chicago Press.

Vaughan, T. A. 1976. Nocturnal behavior of the African false vampire bat (*Cardiodermacor*). *J. Mammal.* 57:227–248.

Vaughan, T. A., J. M. Ryan, and N. J. Czaplewski. 2000. *Mammalogy.* 4th ed. Fort Worth, TX: Saunders College Publishing.

Vickery, W. L., and J. S. Millar. 1984. The energetics of huddling by endotherms. *Oikos* 43:88–93.

Vogel, P. 1976. Energy consumption of European and African shrews. *Acta Theriol.* 21:195–206.

———. 1980. Metabolic levels and biological strategies in shrews. In *Comparative physiology: Primitive mammals,* ed. K. Schmid-Nielsen, L. Bolis, and C. R. Taylor, 170–180. Cambridge: Cambridge Univ. Press.

Vogt, F. D., and G. R. Lynch. 1982. Influence of ambient temperature, nest availability, huddling, and daily torpor on energy expenditure in the white-footed mouse *Peromyscus leucopus. Physiol. Zool.* 55:56–63.

Walsberg, G. E. 1983. Coat color and solar heat gain in animals. *Bioscience* 33: 88–91.

———. 1991. Thermal effects of seasonal coat change in three subarctic mammals. *J. Thermal Biol.* 16:291–296.

Wang, L. C. H. 1978. Energetic and field aspects of mammalian torpor: The Richardson's ground squirrel. In *Strategies in cold: Natural torpidity and thermogenesis,* ed. L. C. H. Wang and J. W. Hudson, 109–145. New York: Academic Press.

———. 1979. Time patterns and metabolic rates of natural torpor in the Richardson's ground squirrel. *Can. J. Zool.* 57:149–155.

Waterman, J. M. 1995. The social organization of the Cape ground squirrel (*Xerus inauris*: Rodentia: Sciuridae). *Ethology* 101:130–147.

Waterman, J. M., and J. D. Roth. 2007. Interspecific associations of Cape ground squirrels with two mongoose species: Benefit or cost? *Behav. Ecol. Sociobiol.* 61:1675–1684.

Webb, P. I., and J. Ellison. 1998. Normothermy, torpor, and arousal in hedgehogs (*Erinaceus europaeus*) from Dunedin. *N. Z. J. Zool.* 25:85–90.

Weigl, P. D., M. A. Steele, L. J. Sherman, J. C. Ha, and T. L. Sharpe. 1989. The ecology of the fox squirrel (*Sciurus niger*) in North Carolina: Implications for survival in the southeast. *Bull. Tall Timbers Res. Stn.* 24:1–93.

Weldon, P. J. 2004. Defensive anointing: Extended chemical phenotype and unorthodox ecology. *Chemoecology* 14:1–4.

Whitaker, J. O. Jr. 1962. *Endogone, Hymenogaster,* and *Melanogaster* as small mammal foods. *Am. Midl. Nat.* 67:152–156.

———. 1963a. Food, habitat and parasites of the woodland jumping mouse in central New York. *J. Mammal.* 44:316–321.

———. 1963b. Food of 120 *Peromyscus leucopus* from Ithaca, New York. *J. Mammal.* 44:418–419.

———. 1995. Food of the big brown bat *Eptesicus fuscus* from maternity colonies in Indiana and Illinois. *Am. Midl. Nat.* 134:346–360.

———. 2004. Prey selection in a temperature zone insectivorous bat community. *J. Mammal.* 85:460–469.

Wiens, F., A. Zitzmann, and N. A. Hussein. 2006. Fast food for slow lorises: Is low metabolism related to secondary compounds in high-energy plant diet? *J. Mammal.* 87:790–798.

Wiley, R. W. 1980. *Neotoma floridana. Mamm. Species* 139:1–7.

Willi, U. B., G. N. Bronner, and P. M. Narins. 2006. Middle ear dynamics in response to seismic stimuli in the Cape golden mole (*Chrysochloris asiatica*). *J. Exp. Biol.* 209:302–313.

Williams, O., and B. A. Finney. 1964. *Endogone*—food for mice. *J. Mammal.* 45: 265–271.

Williams, T. M. 1990. Heat transfer in elephants: Thermal partitioning based on skin temperature profiles. *J. Zool. Lond.* 222:235–245.

Willis, K., M. Horning, D. A. S. Rosen, and A. W. Trites. 2005. Spatial variation of heat flux in Steller sea lions: Evidence for consistent avenues of heat exchange along the body trunk. *J. Exp. Mar. Biol. Ecol.* 315:163–175.

Willmer, P., G. Stone, and I. Johnston. 2000. *Environmental physiology of animals.* Oxford: Blackwell Science.

Wilson, D. E. 1997. *Bats in question: The Smithsonian answer book.* Washington, DC: Smithsonian Institution Press.

Wilson, D. E., and D. M. Reeder, eds. 2005. *Mammal species of the world: A taxonomic and geographic reference.* 3d ed., 2 vols. Baltimore: Johns Hopkins Univ. Press.

Wilz, M., and G. Heldmaier. 2000. Comparison of hibernation, estivation and daily torpor in the edible dormouse, *Glis glis. J. Comp. Physiol. B* 170:511–521.

Wischusen, E. W. 1990. The foraging ecology and natural history of the Philippine flying lemur (*Cynocephalus volans*). Ph.D. diss., Cornell Univ., Ithaca, NY.

Wischusen, E. W., and M. E. Richmond. 1998. Foraging ecology of the Philippine flying lemur (*Cynocephalus volans*). *J. Mammal.* 79:1288–1295.

Withers, P. C., K. C. Richardson, and R. D. Wooller. 1990. Metabolic physiology of euthermic and torpid honey possums, *Tarsipes rostratus. Aust. J. Zool.* 37: 685–693.

Wolff, J. O. 1980. Social organization of the taiga vole (*Microtus xanthognathus*). *Biologist* 62:34–45.

Wolff, J. O., and D. S. Durr. 1986. Winter nesting behavior of *Peromyscus leucopus* and *Peromyscus maniculatus. J. Mammal.* 67:409–412.

Wolff, J. O., and W. Z. Lidicker Jr. 1981. Communal winter nesting and food sharing in taiga voles. *Behav. Ecol. Sociobiol.* 9:237–240.

Wooller, R. D., M. B. Renfree, E. R. Russell, A. Dunning, S. W. Green, and P. Duncan. 1981. Seasonal changes in population of the nectar-feeding marsupial *Tarsipes spencerae* (Marsupialia: Tarsipedidae). *J. Zool. Lond.* 195:267–279.

Wooller, R. D., E. M. Russell, and M. B. Renfree. 1984. Honey possums and their food plants. In *Possums and gliders,* ed. A. P. Smith and I. D. Hume, 439–443. Sydney: Australian Mammal Society.

Wozencraft, W. C. 2005. Order Carnivora. In *Mammal species of the world: A taxonomic and geographic reference,* 3d ed., vol. 1, ed. D. E. Wilson and D. M. Reeder, 532–628. Baltimore: Johns Hopkins Univ. Press.

Wrabetz, M. J. 1980. Nest insulation: A method of evaluation. *Can. J. Zool.* 58: 938–940.

Wroe, S., C. McHenry, and J. Thomason. 2005. Bite club: Comparative bite force in big biting mammals and the prediction of predatory behaviour in fossil taxa. *Proc. R. Soc. B* 272:619–625.

Wroot, A. 1984. Hedgehogs. In *The Encyclopedia of mammals,* ed. D. Macdonald, 751–757. New York: Facts on File.

Wunder, B. A. 1984. Strategies for, and environmental cueing mechanisms of, seasonal changes in thermoregulatory parameters of small mammals. In *Winter ecology of small mammals,* ed. J. F. Merritt, 165–172. Carnegie Museum of Natural History spec. pub. no. 10. Pittsburgh: Carnegie Museum of Natural History.

Wunder, B. A., D. S. Dobkin, and R. D. Gettinger. 1977. Shifts of thermogenesis in the prairie vole (*Microtus ochrogaster*): Strategies for survival in a seasonal environment. *Oecologica* 29:11–26.

Yates, T. L., and D. J. Schmidly. 1978. *Scalopus aquaticus. Mamm. Species* 105:1–4.

Yeager, L. E. 1943. Storing of muskrats and other foods by minks. *J. Mammal.* 24: 100–101.

Yeboah, S., and K. B. Dakwa. 2002. Colony and social structure of the Ghana mole-rat (*Cryptomys zechi,* Matchie) (Rodentia: Bathyergidae). *J. Zool.* 256:85–91.

Zhang, Y. Y., R. Proenca, M. Maffei, M. Barone, L. Leopold, and J. M. Friedman. 1994. Positional cloning of the mouse obese gene and its human homolog. *Nature* 372:425–432

Index

Calomyscidae, 13
Campylorhynchus brunneicapillus, 201
Canada lynx, 251–53
Cane rats, 15
Canidae, 2, 10, 105, 111, 114–17
Canine teeth, 258; of bats, 146; of bush-babies, 94; of carnivores, 10, 105–6; of cats, 107; of fennecs, 116; of herbivores, 61; of hyraxes, 102; of mouse lemurs, 89; of mustelids, 110; of shrew opossums, 137; of vampire bats, 123, 124
Canis: C. latrans, 197; *C. lupus,* 115, 173; *C. lupus dingo,* 115
Cape ground squirrels, 205–7
Capreolus, 225; *C. capreolus,* 225; *C. pygargus,* 225
Capromyidae, 15
Capybaras, 11, 14, 15, 60
Carcophilus harisii, 112
Cardioderma cor, 53, 146
Caribou, 201, 210
Carica papaya, 77
Carnassial teeth, 10, 105–6, 126, 258; of cats, 107; of mustelids, 110; of procyonids, 127
Carnegia gigantea, 82
Carnivores, 10–11, 105–25, 189, 258; canids, 114–17; delayed implantation in, 225; desert-adapted, 196; felids, 106–10; flesh-eating, 106–17; insectivorous, 134–135; marsupials, 139–42; mustelids, 110–14; omnivorous, 10, 126–29; piscivores, 117–22; sanguinivores, 122–25; taxonomy of, 10–11, 105; teeth of, 10, 105–6
Carya, 68
Castor canadensis, 13, 175, 234
Castoridae, 11, 12, 13
Cats, 10, 105, 106–10
Cavernous sinus, 258
Cavies, 11, 15
Caviidae, 11, 15
Cebidae, 9, 91
Cebuella, 89
Cecum, 24, 63, 64, 258; of carnivores, 106; hindgut fermentation in, 96–97; of hyraxes, 102, 103; of lorisiforms, 94; of omnivores, 126; of sengis, 236
Ceiba, 83; *C. pentandra,* 77

Cellulolytic enzymes, 95, 100, 258
Centropogon nigricans, 83
Cerastes: C. cerastes, 109; *C. vipera,* 109
Cercopithecidae, 9, 208
Cervidae, 213
Chaetodipus, 158
Cheetahs, 107
Cheirogaleidae, 9, 89
Cheirogaleus, 89; *C. medius,* 160
Chimarrogale, 40
Chimpanzees, 222
Chinchilla rats, 15
Chinchillidae, 15
Chipmunks, 13, 68, 72, 136, 204
Chironectes minimus, 121
Chiroptera, 7–8, 21, 48–50, 189, 194. *See also* Bats
Chiropterogamy, 77
Chitty hypothesis, 251
Choeronycteris mexicana, 82, 83, 146
Chrysochloridae, 6, 24, 44, 46–47
Chrysochloris asiatica, 48
Cicadas, 95
Cingulata, 21, 29, 194
Circumpolar regions, 111, 115, 173, 258
Civets, 10, 105, 126, 128–29
Clethrionomys. See *Myodes*
Cloaca, 258; of insectivores, 7; of lagomorphs, 16; of marsupials, 221; of monotremes, 221
Coatis, 11, 126, 127
Cold environments, 147–48, 149, 150, 172–91; countercurrent heat exchange in, 174–76; Dehnel's phenomenon in, 179–80, 259; fat tails in, 180–83; hibernation in, 160–71; increased heat production in, 189–91; insulatory changes in, 172–74; maintaining energy balance in, 172; modified size of appendages in, 208–10; reduced activity level in, 175–79; social thermoregulation in, 183–89; survival mechanisms in, 149, 150; torpor in, 159; white pelage in, 210–13
Collocalia, 58
Colocolos, 109, 183
Color dimorphism, 210–14
Colugos, 6, 7, 98–101
Combretum molle, 102
Commensal mammals, 11, 60, 258

Communal nesting, 183–89, 216–18
Concealment from predators, 2; coloration for, 212–13
Condylura cristata, 36–37, 39, 181
Conies. *See* Hyraxes
Convergence, 6–7, 24, 34, 38, 43, 47, 74, 235, 258
Coprophagy, 63–65, 258; in lagomorphs, 15, 63; in mole-rats, 245; in mountain beavers, 63–65; in rodents, 12; in shrews, 63, 65
Coruros, 15, 44
Cougars, 109
Countercurrent heat exchange, 174–76, 199–200, 258
Coyotes, 114, 197
Craseonycteridae, 8
Craseonycteris thonglongyai, 8
Crepuscular mammals, 2, 258
Cricetidae, 13, 159, 194
Critical temperature: lower, 153, 154, 156, 173, 261; upper, 153, 154, 266
Crocidura, 181; *C. russula,* 23, 159; *C. suaveolen,* 159
Crocidurinae, 22, 24, 159
Crocuta, 62
Cryptomys: C. damarensis, 244, 245; *C. hottentotus,* 234; *C. zechi,* 244
Cryptotis, 137
Crytoprocta ferox, 129
Ctenodactylidae, 15
Ctenomyidae, 15, 74–76
Ctenomys, 74; *C. boliviensis,* 75; *C. pearsoni,* 76; *C. talarum,* 74; *C. tucumanus,* 74
Cud, 96
Cuniculidae, 15
Cursorial locomotion, 11, 116, 259
Cynictis pencillatus, 206
Cynocephalidae, 6, 98
Cynocephalus: C. variegates, 98; *C. volans,* 98
Cynomys, 206; *C. ludovicianus,* 117
Cynopterus sphinx, 228
Cystophora cristata, 222

Dactylopsila palpator, 32–33
Daily torpor, 158–59, 259
Dama dama, 240
Damaliscus lunatus, 240

Dassie rats, 15
Dasycercus cristicauda, 135–37, 197
Dasypodidae, 194
Dasyproctidae, 15
Dasypus novemcinctus, 227
Dasyuridae, 105, 139–42, 193–94, 247
Dasyuromorphia, 21, 105, 189, 194
Daubentonia madagascariensis, 32–33
Daubentoniidae, 9
Daubenton's bat, 117, 120–21
Deer mice, 186
Dehnel's phenomenon, 179–80, 259
Dehydration, 197
Delayed development, 225, 228
Delayed fertilization, 223–24
Delayed implantation, 7, 225–28, 259
Dendrohyrax, 16, 102; *D. arboreus,* 102; *D. dorsalis,* 102
Dermopterans, 6–7, 98–101, 234
Desert-adapted mammals, 149–50, 153, 192–207; behavioral avoidance of heat in, 204–7; dietary water intake in, 195–97; estivation in, 158; evaporative cooling in, 197–202; pelage insulation in, 203–4; water conservation in, 193–95
Desmana moschata, 38
Desmans, 6, 24, 36, 38–39
Desmodus, 20; *D. rotundus,* 122–25
Dholes, 114
Diaemus youngi, 122
Diastema, 259; in aye-ayes, 32; in lagomorphs, 15; in rodents, 11, 62
Dicrostonyx, 174, 210; *D. groenlandicus,* 251
Didelphinae, 139, 181
Diet, 2, 3, 19–20; of bats, 7, 21, 48–49; of bushbabies, 93–94; carnivorous, 10–11, 105–25, 258; of desmans, 38; of dormice, 161; of felids, 107–10; of fennecs, 115–16; folivorous, 95–101, 260; frugivorous, 76–82, 260; of gliding possums, 91–93; granivorous, 65–76, 260; gummivorous, 89–95, 260; of hedgehogs, 219; herbivorous, 19, 60–104, 260; of hyraxes, 102; insectivorous, 2, 6–7, 19, 21–59, 134, 261; of kinkajous, 81–82; of mice, 2; of minks, 114; mycophagous, 129–31, 262; myrmecophagous, 28–32, 134,

Hyaenidae, 10
Hydrochaerus hydrochaeris, 11, 60
Hydrochoeridae, 11
Hyenas, 10, 62, 105
Hylobatidae, 9
Hymenoptera, 242
Hyperosmotic urine, 193–94
Hypogeomys antimena, 234
Hypothalamus, 152, 153, 190, 200
Hypothermia, 153, 156, 184, 261; adaptive, 257
Hypsignathus monstrosus, 240–42
Hypsodont dentition, 16, 102, 261
Hyracoidea, 16–17, 101
Hyraxes, 16–17, 63, 101–4
Hystricidae, 15
Hystricomorph rodents, 14–15, 261

Ia io, 145
Ichneumia albicauda, 128
Implantation, 261; delayed, 7, 225–28, 259
Incisors, 261; of aye-ayes, 32; of bats, 146; of bushbabies, 94, 95; of carnivores, 105; of colugos, 99, 100, 101; of herbivores, 61–63; of honey possums, 86; of hyraxes, 102; of insectivores, 22; of lagomorphs, 15–16, 61; of marmosets, 89–90; of mole-rats, 242–44, 246; of mouse lemurs, 89; of pocket gophers, 72–73; of rodents, 11; of shrew opossums, 137; of tuco-tucos, 74; of vampire bats, 123, 124
Incrassated tail, 137–38, 181–83, 261
Incus, 48, 57, 261
Indriidae, 9
Infraorbital foramen, 12, 13, 261
Insectivores, 2, 6–7, 19, 21–59, 134, 261; aerial (bats), 48–59; arboreal, 32–34; delayed implantation in, 225; dentition of, 21–23; desert-adapted, 196; estivation by, 158; protrusile tongue of, 28–32; semiaquatic and fossorial, 34–42; sensory mucous glands of, 34–36; subterranean, 39, 42–48; taxonomy of, 6–7, 21, 24; terrestrial, 24–32; venomous saliva of, 7, 25–28
Insensible water loss, 198, 261
Insulatory changes, 172–74; in arctic foxes, 173; in snowshoe hares, 173–74

Ipomoea, 83
Iridomyrmex detectus, 30
Iteroparity, 247, 261

Jackals, 114
Jackrabbits, 116, 117, 175, 177, 208
Jaguarondis, 109
Jaguars, 109
Jerboa-marsupials, 142
Jerboas, 13, 66, 180, 194, 235
Jumping mice, 13, 19, 169–71

Kangaroo mice, 72, 158
Kangaroo rats, 62, 66, 70, 71, 72, 120, 158, 180, 194, 195, 196, 201–2, 235
Kangaroos, 229, 230
Kannabateomys amblyonyx, 234
Kerivoula papuensis, 143
Kidney structure and function, 193–94; in aquatic mammals, 193; in dasyurid marsupials, 193–94; in desert mammals, 193, 194; in halophyte-eating mammals, 195; in hyraxes, 104; in mulgaras, 197
Kinkajous, 80–82, 126, 127
Kirkia acuminate, 102
Kit foxes, 116–17, 197, 208, 209
Kleiber's law, 155
Koalas, 101
Kobus kob, 240
Kodkods, 109
K-selection, 14, 261
Kultarrs, 142

Lactation, 222, 261; body temperature during, 159; of Cape ground squirrels, 207; of eutherians, 221; of honey possums, 229, 230; of marsupials, 5
Lagomorpha, 189
Lagomorphs, 15–16, 61, 194; caching by, 70; coprophagy in, 15, 63; teeth of, 15–16, 61
Lagorchestes conspicillatus, 195
Lama guanacoe, 203
Lasionycteris noctivagans, 223
Lasiurus borealis, 145
Latitude and body mass, 214–16
Leadbeater's possums, 89, 93
Lek behavior, 240–42, 261
Lemaireocereus, 83

Lemmings, 13, 14, 174, 179, 210, 249–52, 254
Lemmus, 174, 251; *L. lemmus,* 252; *L. sibiricus,* 249; *L. trimucronatus,* 251, 252
Lemur flavus, 81
Lemuridae, 9
Lemurs, 9, 89, 101, 129
Leopardus: L. colocolo, 109; *L. geoffroyi,* 109; *L. guigna,* 109; *L. jacobita,* 109; *L. pajeros,* 109; *L. pardalis,* 109; *L. tigrimus,* 108, 109, 110; *L. wiedii,* 108, 109
Lepilemuridae, 9
Leporidae, 15–16, 194
Leptin, 181
Leptonycteris nivalis, 82–83, 146
Lepus, 62, 210; *L. americanus,* 173, 177, 208, 210, 213, 251–52; *L. arcticus,* 210, 212; *L. californicus,* 116, 175, 177, 208; *L. capensis,* 108; *L. timidus,* 210; *L. townsendii,* 117
Lestoros inca, 137
Life span, 3
Limnogale mergulus, 233
Liomys, 72, 158
Lions, 107
Lipase, 190
Litters: of antechinus, 248; of hedgehogs, 220; intervals between birth of, 222; of mole-rats, 15, 244; size of, 222; of tenrecs, 231, 244
Lophodont dentition, 102, 261
Lorises, 9, 89, 94
Lorisidae, 9
Lower critical temperature, 153, 154, 156, 173, 261
Loxodonta africana, 203
Lycalopex sechurae, 65
Lynx canadensis, 251–53

Macaca, 208
Macroderma gias, 146
Macroglossus, 84, 146; *M. minimus,* 77, 78
Macropodidae, 139, 194
Macropus rufus, 6
Macroscelidea, 6–7, 234, 237
Macroscelides proboscideus, 234
Macroscelididae, 234

Macrotus californicus, 228
Madagascar mongooses, 126, 129
Malacothrix typica, 108
Malagasy civets, 129
Malleus, 48, 57, 261
Mammae, 261; of bushbabies, 94; of guanacos, 203; of kultarrs, 142; of mole-rats, 244, 246
Mammalian species, 1, 3–17. *See also* Small mammals
Mangifera indica, 77
Margays, 108, 109–10, 127
Marmosa, 139, 247; *M. murina,* 139; *M. robinsoni,* 138, 139
Marmosets, 9, 89–91, 130
Marmosops, 139, 247
Marmota, 160; *M. alpina,* 189; *M. flaviventris,* 188; *M. marmota,* 188; *M. monax,* 1, 102, 160, 188, 203–4; *M. olympus,* 188
Marmots, 13, 160, 188, 189
Marsupialia, 5
Marsupials, 4, 5–6, 261; antechinuses, 247; body temperature of, 153; brown fat in, 189; caenolestids, 137; carnivorous, 139–41; compared with eutherians, 5–6; coprophagy in, 63; embryonic diapause in, 228–30; estivation by, 158; heat production in, 153; honey possums, 85–87; kidneys of, 193–94; numbats, 30–32; reproduction of, 221, 222, 228–30; taxonomy of, 5; water opossums, 121–22
Marsupium, 5, 221
Martes, 213; *M. pennanti,* 213, 226–27
Masseter muscle, 261–62; of carnivores, 106; of herbivores, 61; of rodents, 12, 13, 14
Mate guarding, 236
Maternal care, absentee, 237–40
Mating systems, 231, 234; lek behavior, 240–42
Meerkats, 128, 134–35
Megachiropterans, 7–8, 48, 228; frugivorous, 77–79; nectarivorous, 82, 83–84, 146. *See also* Bats
Megaderma: M. lyra, 145, 146; *M. spasma,* 146
Megadermatidae, 53, 143, 145, 146
Melaleuca, 84

Mustela, 210, 216, 251; M. erminea, 211, 213; M. frenata, 211, 227; M. nivalis, 110–12, 227
Mustelidae, 10, 11, 105, 106, 110–14, 225, 226, 233
Mycophagy, 129–31, 262
Mycorrhiza, 131, 262
Myocastoridae, 15
Myodes: M. gapperi, 14, 179; M. rutilus, 190
Myomorph rodents, 13–14, 262
Myotis, 223; M. adversus, 117, 120, 121; M. albescens, 228; M. brandti, 3; M. daubentonii, 117, 120–21; M. lucifugus, 49, 167, 223; M. sodalis, 49; M. vivesi, 117, 119–20, 193, 194
Myremecobius fasciatus, 31
Myrmecobiidae, 30
Myrmecophagous mammals, 28–32, 134, 262
Myrtillocactus, 83

Nandiniidae, 10
Nannospalax, 46
Napaeozapus, 19, 169; N. insignis, 169, 171
Nectarivores, 82–88, 262; bats, 82–85, 146; feather-tailed gliders, 87–88; honey possums, 85–87; tongue of, 82–85
Nectogale, 40; N. elegans, 40
Neomys, 40; N. anomalous, 25; N. fodiens, 23, 25
Neotoma, 216; N. floridana, 70; N. lepida, 195
Neovison vison, 38, 111, 114
Nesomyidae, 13
Nests: communal, 183–89, 216–18; thermal capacity of, 186–88
Newtonia, 94
Ningaui, 21, 86, 139–41; N. ridei, 194; N. timealeyi, 6; N. yvonneae, 140
Ningauis, 6, 21, 86, 139–41
Noctilio: N. albiventris, 118–19; N. leporinus, 117–19, 120, 121
Noctilionidae, 117, 143
Nocturnal mammals, 2, 262; aye-ayes, 32; bushbabies, 94; caenolestids, 137; colugos, 100; felids, 107; fennecs, 115; flying squirrels, 218; kinkajous, 81;

lemurs, 89; mice, 13, 158; monito del montes, 183; mouse opossums, 139; ningauis, 139; pen-tailed treeshrews, 238; ringtails, 127; weasels, 112
Nonshivering thermogenesis, 148, 153, 168, 189, 190–91, 218, 262
Noolbengers, 85–87
Norepinephrine, 190, 262
Notomys: N. alexis, 67, 194; N. cervinus, 194
Notoryctemorphia, 21
Notoryctidae, 44, 47
Numbats, 29, 30–32
Nutrias, 15
Nyctalus lasiopterus, 145
Nyctea scandiaca, 251
Nycteridae, 53, 143, 145
Nycteris thebaica, 53
Nycticebus coucang, 89

Occlusal surfaces of teeth, 262; in herbivores, 63; in insectivores, 22
Ocelots, 109
Ochotona, 15, 70; O. dauurica, 71; O. pallasi, 71; O. princeps, 70–71
Ochotonidae, 15–16
Ochroma lagopus, 77
Octodontidae, 15, 194
Odobenidae, 11
Olingos, 81, 126, 127
Omasum, 96, 262
Omnivores, 10, 105, 116, 126–46, 262; carnivorous, 126–29; dietary nonconformists, 134–42; mycophagous, 129–31
Oncillas, 108, 109, 110
Ondatra zibethicus, 185
Onychomys: O. leucogaster, 117, 135, 136; O. torridus, 196
Onychonycteris finneyi, 50
Opossums, 21, 121–22, 126, 181–82, 222
Opuntia, 195
Orangutans, 222
Ornithorhynchidae, 4–5, 34
Ornithorhynchus anatinus, 5, 35
Oryzorictes, 233
Otariidae, 11
Otocyon megalotis, 115
Otolemur crassicaudatus, 93

Reproductive variations, 222; delayed development, 225, 228; delayed fertilization, 223–24; delayed implantation, 7, 225–28, 259; embryonic diapause, 228–30, 259
Respiration rate, in torpor, 156–57
Respiratory alkalosis, 199
Respiratory heat exchange, 200–202
Rete mirabile, 174, 264
Reticulum, 96, 264
Rhinolophidae, 159, 223
Rhinolophus rouxii, 228
Rhinonycteris aurantius, 202
Rhynchocyon, 237; *R. udzungwensis,* 235
Rhyncholestes raphanurus, 137, 138, 182
Ringtails, 65, 126, 127, 128
Rock badgers, 16–17, 63, 101–4
Rodents *(Rodentia),* 11–15, 60–61, 189; adaptations of, 11, 60; caching by, 67, 68–70; cheek pouches of, 71–72; coprophagy in, 12, 63; delayed implantation in, 225; desert, 204–5; diet of, 12, 13, 14, 19; estivation by, 158; fossorial, 74; granivorous, 19, 65, 67, 205; hystricomorph, 14–15; insectivorous, 19, 134, 135, 150; monogamous, 234; myomorph, 13–14; population cycles of, 251; r-selected vs. K-selected species of, 14; sciuromorph, 12–13; size of, 11, 60; species of, 60; stomach of, 96; teeth of, 11–12; torpor by, 158, 159; urine-concentrating ability of, 194
Roe deer, 225–26
Rostrum, 264; of canids, 115; of felids, 107; of shrews, 24; of vampire bats, 123
Rousettus, 58; *R. aegyptiacus,* 79
r-selection, 14, 264
Rumen, 96, 264
Ruminate, 95–96, 98, 264

Saccolaimus peli, 53
Saccopteryx binineata, 145
Salanoia concolor, 129
Saliva: self-anointing of hedgehogs, 220; spreading of, 200; of vampire bat, 124; venomous, 7, 25–28
Salivary amylase, 96, 264
Saltatorial locomotion, 16, 142, 234, 264

Sand cats, 108, 109
"Sand puppies," 242, 243
Sand rats, 194, 195–96
Sanguinivores, 105, 122–25, 264
Sarcophilus harrisii, 139
Scalopus aquaticus, 44–45
Scandentia, 6–7, 238
Scansorial mammals, 11, 264
Scarabaeidae, 135
Sciuridae, 12, 96, 194
Sciuromorph rodents, 12–13, 264
Sciurus, 68; *S. carolinensis,* 68, 69, 121, 205, 215, 235; *S. niger,* 68, 69, 215; *S. vulgaris,* 130
Sclerophyll environments, 92, 264
Scutisorex somereni, 131–34
Seals, 11, 222, 225
Seasonal color dimorphism, 210–14
Sechuran foxes, 65
Seismic sensitivity, 46–48
Semelparity, 246–48, 264
Semiaquatic mammals, 42; insectivores, 34–40; web-footed tenrecs, 233
Sengis, 6, 7, 231, 234–37
Setifer, 233; *S. setosus,* 233
Set point, 152–53, 200, 264
Sheep, 173
Shivering, 148, 153, 189
Shrew-moles, 6, 24
Shrew opossums, 21, 137, 138, 182
Shrews, 21; activity in winter, 177–78; aggression of, 27–28; brown fat in, 191; caching by, 27, 67, 177; coprophagy in, 63, 65; Dehnel's phenomenon in, 179; diet of, 21, 23, 24–25; geographic distribution of, 24, 25; hero (armored), 131–34; incrassated tail of, 181; life span of, 3; nonshivering thermogenesis in, 191; skull of, 22–23; South American, 137–38; taxonomy of, 6–7, 24; teeth of, 21–22, 24; torpor by, 159–60; underwater sniffing of, 40–42, 43; venomous saliva of, 7, 25–28; water, 23, 25, 40–42, 43
Siamangs, 9
Sifakas, 9
Simplex uterus, 9, 264–65
Skunks, 11, 34, 105, 159, 183–84, 226, 227
Sloth bears, 134

Sloths, 29, 101, 175

Small mammals: advantages and disadvantages of, 2–4; evolution of, 3–4; life spans of, 3; tails of, 180; weight definition of, 1–2

Sminthopsis, 139; *S. crassicaudata,* 141; *S. longicaudata,* 140, 141; *S. murina,* 141

Snowshoe hares, 173–74, 177, 208, 210, 251–54

Snowy owls, 251

Social thermoregulation, 183–89

Solenodon paradoxus, 25, 26

Solenodons, 6, 22, 24; venomous saliva of, 7, 25–27

Solenodontidae, 6, 24

Sorex, 40; *S. alaskanus,* 40; *S. araneus,* 25, 177, 180; *S. bendirii,* 40; *S. caecutiens,* 23; *S. cinereus,* 180, 191; *S. hoyi,* 24; *S. minutus,* 25; *S. palustris,* 23, 40–43

Soricidae, 6, 24, 40, 137

Soricinae, 24, 25, 159

Soricomorpha, 6–7, 21, 24, 189

Spalacidae, 13, 46

Spalax: S. ehrenbergi, 46; *S. microphthalmus,* 70

Specialists, 251, 265; bats, 142, 144; mustelids, 110; yellow-bellied gliders, 92

Speothos venaticus, 209

Spermophilus: S. lateralis, 113, 161; *S. parryii,* 165; *S. richardsonii,* 168; *S. saturatus,* 130; *S. tridecemlineatus,* 117

Spilogale: S. gracilis, 226; *S. putorius,* 226, 227

Spines: of hedgehogs, 218–20, 221; of tenrecs, 233

Spiny rats, 15

Springhares, 12

Squirrel gliders, 92–93

Squirrels, 12–13, 62; Bergmann's rule for, 215–16; caching by, 13, 67, 68–69; communal nesting of, 188–89, 216–18; desert-adapted, 205–7; hibernation of, 165–66, 168

Stapedius muscle of bats, 57

Stapes, 57, 265

Star-nosed moles, 36–39, 181

Steatornis caripensis, 58

Stomach: of dormouse, 161; of greater bulldog bat, 119; of herbivores, 96–97; of honey possum, 87; of omnivores, 126; of vampire bat, 123, 124

Strepsirhini, 9, 94

Sublingua of bushbabies, 94

Submaxillary glands, 27, 200, 265

Subnivean spaces, 2, 112, 174, 265; nesting in, 184, 185

Subterranean mammals, 2, 42–44, 202, 265; insectivores, 39, 42–48; seismic sensitivity of, 46–48; weasels, 216

Sugar gliders, 92

Suncus: S. etruscus, 24, 159; *S. murinus,* 24

Surface area-to-volume ratio, 151, 197, 214–16

Suricata suricatta, 128, 134, 206

Sus scrofa, 126

Sweat glands, 197, 198, 200; apocrine, 198, 257; eccrine, 198, 259

Sweating, 198–99

Swift foxes, 116–17

Syconycteris, 84, 146; *S. australis,* 84–85

Sylvilagus, 210; *S. audubonii,* 117

Syzigium, 84

Tachyglossidae, 5

Tachyglossus aculeatus, 5, 29–30, 156

Tadarida brasiliensis, 49, 57, 58

Taiga voles, 186, 187

Talpa europaea, 36, 45, 67

Talpidae, 6, 24, 36, 44, 47

Tamarins, 9

Tamiasciurus: T. douglasii, 68; *T. hudsonicus,* 68, 69

Tamias striatus, 30, 68, 72, 136

Tarsiers, 9, 86

Tarsiidae, 9

Tarsipes rostratus, 85–87, 229

Tasmanian devils, 136, 139

Taxidea taxus, 197

Taxonomy, 4–17

Teeth, 20; of bats, 146; brachyodont, 9, 102, 258; bunodont, 9, 126, 258; of bushbabies, 94, 95; canine, 258; carnassial, 10, 105–6, 126, 258; of carnivores, 10, 105–6; of colugos, 99, 100, 101; cusps of, 21, 126, 259; of dasyurids, 141; diastema between, 11, 15,

Joseph F. Merritt is senior mammalogist with the Illinois Natural History Survey, University of Illinois. He is the former director of Powdermill Biological Station of the Carnegie Museum of Natural History and served as Distinguished Visiting Professor at the U.S. Air Force Academy in Colorado Springs, Colorado, during the academic year 2004 to 2005. Dr. Merritt is a physiological ecologist and functional morphologist specializing in adaptations of mammals to cold. He is the author of *Guide to the Mammals of Pennsylvania,* published by the University of Pittsburgh Press, and coauthor of the college textbook *Mammalogy: Adaptation, Diversity, Ecology,* published by the Johns Hopkins University Press. He is also editor of several technical monographs on specific taxa of mammals. Dr. Merritt has served on the Publications Committee of the *American Society of Mammalogists* since 1990 and is currently associate editor of the *Journal of Mammalogy* and *Acta Theriologica.* He teaches mammalogy at the University of Colorado Mountain Research Station, as well as courses in mammalian ecology and winter ecology at the Adirondack Ecological Center, SUNY College of Environmental Science and Forestry.